A GUIDE TO THE

2009 IRC® Wood Wall Bracing Provisions

A Guide to the 2009 IRC Wood Wall Bracing Provisions

ISBN: 978-1-58001-623-0

Copyright © 2009

By

International Code Council Inc.	and	APA – The Engineered Wood Association
500 New Jersey Avenue, NW, 6th Floor		7011 South 19th Street
Washington, DC 20001		Tacoma, Washington 98466

The information contained in this guide is believed to be accurate; however, it is being provided for informational purposes only. Publication of this document by the ICC should not be construed as the ICC or the APA engaging in or rendering engineering, legal or other professional services. Use of the information contained in this book should not be considered by the user as a substitute for the advice of a registered professional engineer, attorney or other professional. If such advice is required, it should be sought through the services of a registered professional engineer, licensed attorney or other professional.

First printing: October 2009

Printed in the United States of America

Table of Contents

Preface

This illustrated guide was developed to help building designers, builders, building officials and others using the code in the application of the lateral bracing requirements of the 2009 International Residential Code® (IRC). While bracing is just one of many important factors to consider when designing, performing plan review, building, or inspecting a structure, it is a common source of confusion and misapplication. The authors of this publication worked closely with the International Code Council® (ICC) staff to identify and explain the key elements of bracing and to demystify the prescriptive bracing provisions of the IRC.

Some of this book's content has been adopted from *A Guide to the 2006 IRC Wood Wall Bracing Provisions*, published to support the 2006 IRC; however, this edition has been rewritten extensively to reflect the 2009 bracing provisions that are significantly different in organization, scope and detail from the previous version of the code.

As in the 2006 guide, basic concepts of the forces acting on buildings, historical perspective, and the correct application of the IRC bracing provisions are presented in a logical format. Background information, examples, specific applications and suggested solutions can be found throughout the book. The final chapter of the book is dedicated to whole house applications and includes several residential building plans in various wind and seismic regions that have been analyzed for proper wall bracing application. Other aspects of the structure that work in conjunction with bracing to form a safe building, such as diaphragms, wall cladding and inter-element connections, are addressed briefly as necessary to enhance the bracing discussion.

Although written to explain the prescriptive bracing provisions of the 2009 IRC, the information presented in this book may also be applied to those bracing provisions of the International Building Code® (IBC) Section 2308 (Conventional Light Frame Construction) that are similar to the IRC. The provisions of the IRC and IBC have evolved differently, but the underlying concepts and theory of bracing are the same.

The ultimate goal of this guide is to promote the accurate understanding and the correct application of the code, resulting in safer buildings and communities; a goal shared by the ICC and *APA – The Engineered Wood Association*.

This book was authored by two wall bracing experts from APA:

EDWARD KEITH, P.E., is the Senior Engineer for the APA Technical Services Division. With over 25 years experience in wood engineering, product development and building code development, he has served on numerous national committees, including the SBCCI Standards for Hurricane Resistant Residential Construction (SSTD-10) and Seismic Resistant Construction (SSTD-13). Keith is currently a member of the ICC Ad Hoc Committee on Wall Bracing. He is registered in the states of Florida and Washington.

GREG BATES is an APA Engineered Wood Specialist. His responsibilities include supporting the proper application of engineered wood products for efficient and durable building solutions for residential and commercial construction. Bates regularly works with builders, designers and code officials in wall bracing training and practical implementation of the code in the field. He received his B.S. degree in Industrial Technology from the University of Wisconsin – Stout.

ICC staff members who contributed to the book include *JOHN HENRY, P.E.*, Principal Staff Engineer, *GARY NELSON, P.E.*, Senior Staff Engineer, and *HAMID NADERI, P.E.*, Vice President of Product Development. Their technical review, evaluation, analysis, suggestions, clarifications, interpretations, direction, bright ideas and good humor were especially appreciated in light of the short production schedule for this edition. Their input was invaluable to the process and the quality of the end product.

The ICC Technical Services Department also deserves recognition for providing technical reviews of the content and working behind the scenes to keep this publication on track. Special thanks to *LARRY FRANKS, P.E.*, for providing the authors with the resources needed to complete this book.

The authors would have been lost without the efforts of the APA Market Communications team, which developed the figures, edited the text, and designed the pages of this book: *KELLY DEVLIN*, *ANDREW STERNARD*, *MICHAEL MARTIN* and *MARILYN LEMOINE*.

And last, but certainly not the least, ICC and APA would like to express our gratitude to those from other industry organizations who provided their invaluable time and expertise to thoroughly reviewing and improving this publication. Our thanks go out to *GARY EHRLICH, P.E.*, Program Manager, Structural Codes & Standards, of the National Association of Home Builders, *SANDRA HYDE, E.I.T.* of Weyerhaeuser Company, and other individuals from these companies and organizations that contributed.

CHAPTER 1

Understanding the Lateral Forces that Act on a House

All buildings, regardless of size or location, must be designed to safely resist the structural loads anticipated during their lifetime. These loads can be divided into two categories: *vertical loads* and *lateral loads*.

Vertical loads act in the "up" or "down" direction. In most cases the "down" loads are caused by gravity. These loads are the obvious ones: the weight of the building itself (dead load), the weight of everything in the building (live load), and environmental loads, such as those from snow. An example of an "up" load is wind uplift. The "up" load path is the same as the "down" load path except the load is acting in the "up" direction. These loads are easy to understand and typical construction practice has evolved into an efficient system that does a good job of accommodating them. Generally speaking, builders in high wind areas are as comfortable installing uplift straps as they are placing beams on cripple studs.

Because downward loads are always present (due to gravity), any deficiencies in the vertical load path are almost immediately apparent due to structural instability. For example, a beam with support at only one end is not going to stand up during construction because it would fall down at the other end.

The real challenge lies not with the vertical loads, but rather with the "sideways" loads, or, as they are referred to in the design community, *lateral loads*. Lateral loads act in a direction parallel to the ground. Most often the result of wind or seismic (earthquake) forces, lateral loads can cause structures to bend and sway, collapse, or even – if the structure is not well attached to the foundation – roll over.

A wood beam carrying an excessive vertical load may creak, groan, split or deflect over time, warning that repair may be necessary to prevent failure. Because the wind and seismic forces that result in lateral loads are sudden and infrequent, there are no such warning indicators of a pending failure.

In every region of the country, lateral load resistance – an essential part of which is wall bracing – has to be planned during design and built into the structure during construction. While this is especially important in regions susceptible to strong wind and seismic forces, the provisions of the International Residential Code (IRC) make lateral load resistance an important consideration in every part of the country. The IRC prescriptively requires specific building elements to resist lateral forces for all structures within its scope.

Wood-frame construction makes it easy for building professionals to construct strong, attractive, and durable structures that meet code requirements and assure good performance during wind events and earthquakes.

When preparing construction documents to meet the code seismic or wind requirements, it is important to understand how lateral loads act on wood framing systems and how construction detailing and fasteners affect the ultimate lateral performance of a structure. Builders, designers and building officials can use these requirements to ensure strength, quality, and safety in both residential and non-residential buildings. Certainly, a better understanding of these requirements will ensure fewer mistakes in design and plan review, as well as in construction.

Wind forces

During a wind event, wind pushes against one wall while pulling on the opposite wall, as demonstrated in *FIGURE 1.1*. Because the two walls receiving wind pressures – the receiving walls – push and pull the roof in the same direction as the wind, the walls on the other two sides of the structure – the bracing walls – must restrain the roof from moving. When the wind is in the perpendicular direction, the walls change roles: walls that previously restrained the roof now receive the wind pressures, and walls that previously received the wind pressures now must restrain the roof. Thus, all walls must be strong enough to resist the wind forces that push against the home, regardless of whether they must act as a receiving wall or a restraining wall.

FIGURE 1.1

Wind forces acting on a structure

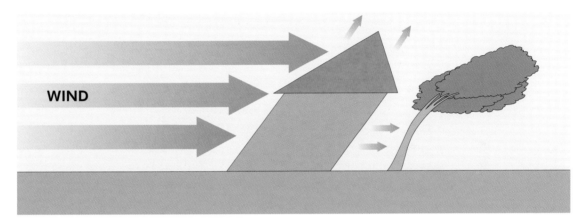

WIND

The IRC wind provisions apply to wind speeds less than 110 miles per hour (mph). In hurricane-prone regions, however, this upper limit is reduced to 100 mph. In areas with wind speeds 110 mph and over – or 100 mph and over in hurricane-prone regions – residential structures must be designed to resist wind loads in accordance with the International Building Code (IBC), or other referenced standards as discussed in *CHAPTER 4* of this publication.

In addition to the basic wind speed, the 2009 IRC requires identification of the Wind Exposure Category of the building site. Wind Exposure Category is explained in Section R301.2.1.4 of the IRC and *CHAPTER 4*. The IRC wind bracing table, IRC Table R602.10.1.2(1), is based on Exposure Category B. For Exposure Categories C and D, bracing increases up to 70 percent are listed in Footnote b. Knowledge of the Wind Exposure Category is also necessary if portions of the structure fall outside the prescriptive scope of the IRC and an engineered design is required. The Wind Exposure Category is also necessary to determine the component and cladding load requirements of IRC Section R301.2.1.

The proper selection of wall sheathing products is essential to ensure the exterior wall assembly has the capacity to resist component and cladding wind pressure and suction forces when acting as the receiving wall. IRC Table R602.3(3) addresses the proper selection of wood structural panel sheathing based on the design wind speed and Exposure Category.

The IRC contains wind maps for all regions of the country in IRC Figures R301.2(4). Note that in hurricane-prone regions, such as Florida, the application of the IRC wind provisions are reduced from wind speeds less than 110 mph to less than 100 mph. A sample wind map is shown in *FIGURE 1.2*.

FIGURE 1.2
Example of partial basic design wind speed map. Adapted from IRC Figure R301.2(4).
Wind speeds shown in miles per hour

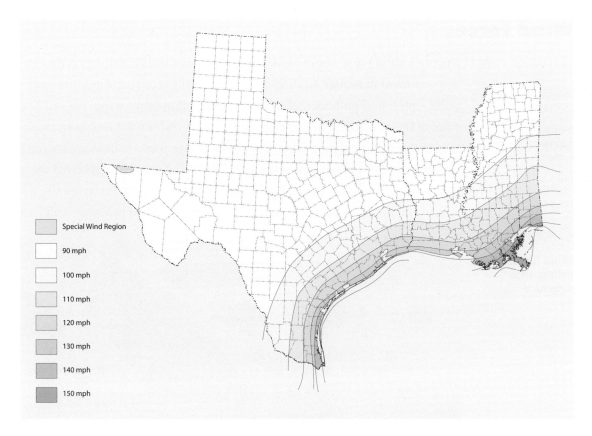

Special Wind Region

90 mph

100 mph

110 mph

120 mph

130 mph

140 mph

150 mph

Seismic forces

Seismic forces are generated by ground motions during an earthquake event. The ground motion causes the structure's mass to accelerate back and forth, up and down. This acceleration causes forces to develop within the structure in locations where the structure's mass is concentrated (Newton's Second Law: Force = Mass x Acceleration). Essentially, the seismic ground motion moves the foundation (*acceleration*), while inertia (*mass* of the structure) attempts to resist this motion. Instead of mass, building codes use seismic weight to determine seismic forces. The seismic weight multiplied by an acceleration expressed as a fraction of the earth's gravity produces the seismic force. Because seismic forces are directly proportional to the weight (mass) of the structure, IRC Section R301.2.2.2.1 imposes limits on weights of materials based on Seismic Design Category (SDC). The seismic weight of the structure is generally concentrated at the floors and roof of the structure. This can be seen in *FIGURE 1.3*.

FIGURE 1.3
Earthquake forces acting on a structure
Vertical (upward forces not shown for clarity)

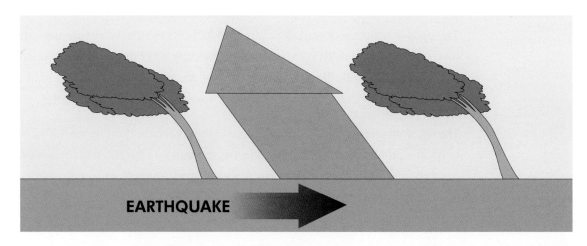

EARTHQUAKE

Similar to the wind maps discussed previously, the IRC publishes earthquake maps that display the various Seismic Design Categories for regions of the country. These maps are located in IRC Figures R301.2(2) (excerpted in **FIGURE 1.4**).

FIGURE 1.4

Example of Seismic Design Category map

Adapted from IRC Figure R301.2(2). Western U.S. shown (all of U.S. not shown for clarity)

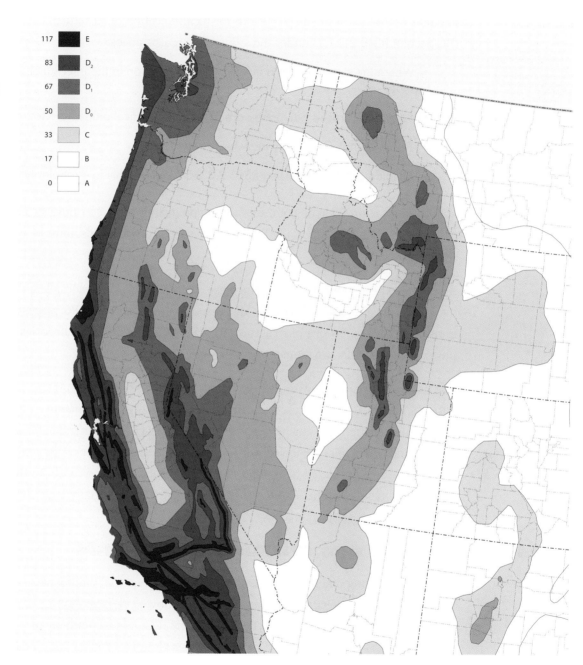

The use of Seismic Design Categories is a simplified means of determining the potential hazard of an earthquake in a region. These categories range from Seismic Design Category (SDC) A to E, with A being the lowest hazard and D_2 the highest covered by the IRC.

For SDC E, the IRC contains provisions for reclassification of structures from SDC E to D_2 (see IRC Section R301.2.2.1.2) under the following circumstances:

- Where the structure is in a SDC D region based on in the International Building Code maps, or

- All exterior braced wall panels are in the same vertical plane from foundation to roof, and

- Floors do not cantilever over exterior walls, and

- Building meets IRC Section R301.2.2.2.5 requirements for "regularity"

For SDC E structures that do not meet these requirements, the prescriptive structural requirements of the IRC cannot be used and the IBC, or other referenced standards, must be used to design the structure.

As can be expected, when the SDC increases from A to D_2, the bracing length and panel attachment requirements also increase. As discussed in **CHAPTER 4**, the IRC places additional limitations on irregular portions of structures in higher Seismic Design Categories, as they may have physical features that are more vulnerable to seismic forces.

The wind and seismic maps in IRC Figures R301.2(4) and R301.2(2) respectively (illustrated in **FIGURES 1.2** and **1.4**) can be used to determine the bracing requirements for a given structure in a particular location. In some regions of the country where the lateral loads for the design-level event (wind or seismic) are lower, minimal bracing is required for residential-type structures. Other areas are considered high wind regions, seismic regions, or both. In these regions, additional wall bracing will be required to accommodate the potentially larger lateral loads.

Determining wind and seismic requirements

Your local building officials may have already determined the wind or seismic requirements (and other design criteria for a region, including snow load, frost depth, termite danger, flood hazard, etc.). When a local jurisdiction adopts the IRC, it is recommended that they fill out IRC Table R301.2(1) for *Climatic and Geographic Design Criteria*. Contact your local building official and ask for a copy of this completed table. (See **TABLE 4.3**.)

What is the lateral load path?

It is very important to understand the concept of a lateral load path because it helps make sense of the prescriptive requirements in the IRC. In short, the lateral load path is simply the path that the lateral or horizontal load takes as it passes through the structure, including components and connections, on its way to the ground. *FIGURE 1.5* shows the critical parts of the lateral load path. Vertical loads follow a similar load path, moving through other structural components on their way to the ground.

The lateral load path for wind loads is simpler to visualize than the load path for seismic loads. *FIGURE 1.5* provides a basic example of the lateral load path resulting from wind loading. The load is shown acting on a windward receiving wall, and its subsequent load path through the building. For simplicity, the suction pressure on the leeward receiving wall is not included in this illustration.

FIGURE 1.5

Critical parts and flow of the load path:

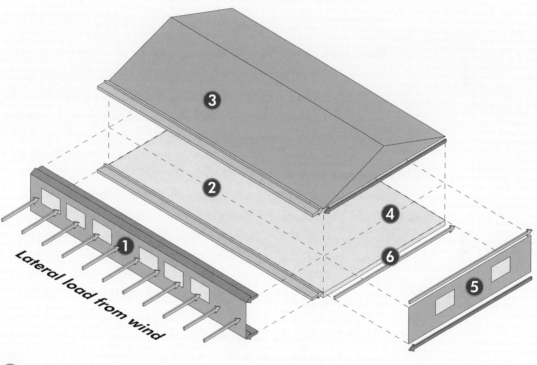

① Windward receiving wall carries load to foundation at bottom of wall and roof diaphragm at top of wall

② Connections at bottom and top of wall transfer these loads into the foundation and diaphragm

③ Roof or floor diaphragm carries load to bracing walls

④ Connections between roof/floor and wall transfers load from diaphragm to bracing walls

⑤ Bracing wall carries load from diaphragm to foundation

⑥ Transfer of loads from the shear walls to the foundation

When discussing the lateral load path, it is helpful to be familiar with commonly used terminology, such as diaphragms, braced wall panels, and shear walls. Section R202 of the IRC defines them as follows:

> **Braced wall line:** *A straight line through the building plan that represents the location of the lateral resistance provided by the wall bracing.*

> **Braced wall panel:** *A full-height section of wall constructed to resist in-plane shear loads through interaction of framing members sheathing material and anchors. The panel's length meets the requirements of its particular bracing method, and contributes toward the total amount of bracing required along its braced wall line in accordance with Section R602.10.1.*

> **Diaphragm:** *A horizontal or nearly horizontal system acting to transmit lateral forces to the vertical resisting elements. When the term "diaphragm" is used, it includes horizontal bracing systems.*

> **Shear wall:** *A general term for walls that are designed and constructed to resist racking from seismic and wind by use of masonry, concrete, cold-formed steel or wood framing in accordance with Chapter 6 of this code and the associated limitations in Section R301.2 of this code.* (Editorial note: this term is most often used in engineered design using the IBC and other appropriate referenced documents.)

In the IRC, diaphragms are simply roofs or floors. Due to the prescriptive nature of the IRC, the diaphragm is "defined" by the minimum thicknesses of roof and floor sheathing (as provided in IRC Table R503.2.1.1(1)) along with the minimum nailing required by IRC Table R602.3(1). Roof and floor diaphragms built to these specifications are deemed to provide sufficient capacity for the loads and exposures covered by the IRC for residential-type structures.

Shear walls and braced wall panels do the same thing: they resist racking from lateral loads. In this book, "shear wall" will be used to refer to an engineered wall segment designed in accordance with the IBC or referenced standards, and "braced wall" or "braced wall panel" will be used to refer to a wall segment constructed in accordance with the prescriptive bracing provisions of the IRC. Further details about the differences between shear walls and braced wall panels are presented in **CHAPTER 2**.

In the vertical load path, all structural members carrying gravity load (e.g., rafters, trusses and joists) must bear on members below (e.g., posts, cripple walls, beams and headers) designed to carry that load. Each of these members must bear on others until the load is transferred through the foundation and into the ground. The vertical load path is fairly easy to understand. Just like a child's first set of building blocks, one block is stacked on another. Due to the force of gravity, vertical loads are always present and it is relatively easy to identify the load path, as shown in **FIGURE 1.6**.

FIGURE 1.6

Example of vertical load path

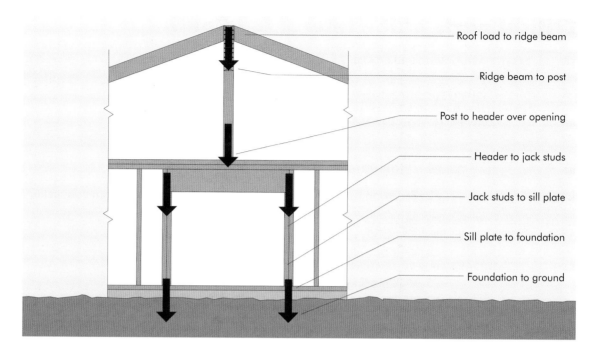

Roof load to ridge beam

Ridge beam to post

Post to header over opening

Header to jack studs

Jack studs to sill plate

Sill plate to foundation

Foundation to ground

While the objective of the lateral load path (to get the applied loads into the ground) is exactly the same as the vertical load path, the actual path it takes is not always obvious. Another key difference is that connections are even more important in the lateral load path. Unlike the vertical load path, in which gravity causes many members to bear on each other, there is nothing holding the different components together in the lateral load path unless the builder makes a connection using nails, straps or framing anchors. The location and requirement for these connections is not always obvious.

Also, unlike vertical loads, lateral loads are intermittent, as high winds and/or earthquakes ("design event") are relatively uncommon occurrences. When these lateral load events do occur, the lateral load path must be in place with each element and connection properly sized and constructed to resist these transient loads.

Critical parts of the lateral load path

As shown in *FIGURE 1.5*, there are six critical parts in the load path for a simple rectangular structure. This section identifies these six parts and explains how the wind and/or seismic loads are distributed through a simple rectangular structure and into the foundation. The critical parts of the lateral load path are:

1. The receiving wall

2. Receiving wall-to-foundation/receiving wall-to-diaphragm connection

3. Floor and roof diaphragm

4. Roof-to-wall/wall-to-wall connections

5. Wall bracing

6. Wall bracing-to-foundation connections

The numbered areas in *FIGURE 1.5* do not necessarily relate to the load path sequence, but are used to differentiate each part, and correlate the following disaster photos and discussion to the theoretical load path concept. Each part of the load path is critical. If any one element in the load path fails, the structure typically fails. Like a chain, the lateral load path is only as strong as the weakest link. While this discussion focuses on a simple rectangular structure, the principles also apply for multi-story structures and T- and L-shaped structures (discussed in *CHAPTER 11*).

1. The receiving wall

The receiving walls are perpendicular to the direction of the lateral load. These walls receive or catch the wind like the sail of a sailboat. The walls themselves must be capable of resisting the loads – for example, both positive (pressure) and negative (suction) wind loads – in order to transfer them to the next link in the load path.

Note that while the receiving wall is under positive pressure, the opposite wall and perpendicular walls are subjected to suction pressures. These suction pressures can pull off poorly attached wall cladding, siding, windows and doors. This loss of nonstructural components is especially critical during high wind events, as it is a failure of the structure's weatherproofing system. High wind events are often accompanied by rain, and while ensuing water damage may not cause a failure of the structural system, it can lead to the total destruction of the contents of the house.

Some examples of wind related failures at the receiving wall due to its inability to resist the applied loads are shown in *FIGURES 1.7* through *1.9*.

Unlike wind, an earthquake acts on the entire structure and not just the receiving walls. The receiving walls, in the case of a seismic event, are not exposed to pressure, but contribute to the seismic weight that is spread throughout the structure. Assuming the receiving walls themselves can withstand the force of the lateral load, the receiving walls transfer this load to the next component in the load path.

FIGURE 1.7

Wall covering is an essential part of the first step of the load path for wind. The wall studs can be seen behind the failed wall covering system. The failure could have been due to various reasons. Approved wall coverings installed per code would have most likely been able to withstand the pressure of the wind. (Photo taken after windstorm in Evansville, Indiana.)

FIGURE 1.8

Not all wall coverings are capable of resisting code-required wind pressures (see IRC Table R301.2(2)). This house was subjected to an 85 mph wind. Failure could have been due to multiple issues, including improper installation or inadequate field-applied connection between the end wall and gable end. (Photo taken after windstorm in Evansville, Indiana.)

FIGURE 1.9

A partial failure of the wall covering system. Note also the house in the background, to the right, which had a similar failure. (Photo taken after Hurricane Katrina in Gulfport, Mississippi, three-to-four miles inland, where wind speeds were within the scope of IRC.)

2. Receiving wall-to-foundation/receiving wall-to-diaphragm connection

The receiving walls must be properly attached at the base and at the top to adequately transfer the forces into the foundation (base attachment via anchor bolts per IRC Section R403.1.6 or attachment to floor framing per IRC Table R602.3(1)) and into the roof diaphragm (attachment per IRC Table R602.3(1)). The proper connection between structural elements of the load path is just as important as the proper selection and detailing of the elements themselves and cannot be over emphasized. Great attention to connection detail is very important in areas of high wind or moderate-to-high seismic force.

3. Floor and roof diaphragm

The floor and roof sheathing form the diaphragms which transfer loads from the receiving wall to the bracing walls. Floor and roof diaphragm failures are rare. When a roof system failure does occur, it is typically due to inadequate attachment of the roof sheathing. This often occurs in roof areas vulnerable to high wind pressure, such as gable ends and roof overhangs, as shown in **FIGURE 1.10**.

FIGURE 1.10

The sheathing was not fastened properly to the roof framing. The loss of sheathing compromises the strength of the roof diaphragm. (Photo taken after a tornado in Southwestern Missouri.)

4. Roof-to-wall/wall-to-wall connections

A proper connection from the roof to the walls below and/or walls to the walls below (in the case of multi-story structures) is critical to the load path and is a common failure point. **FIGURES 1.11** through **1.14** demonstrate this failure mode. As the walls support the roof for both vertical and lateral loads, such a failure can lead to a partial or total building collapse, as illustrated in the figures.

For roof-to-wall connections, it is important to realize that the roof diaphragm actually begins and ends at the exterior wall lines and the connection between the roof and wall is critical to an effective load path (see sidebar.) Connections between the walls and floor are specified in IRC Table R602.3(1). These attachment requirements are minimum connections and are deemed to be sufficient to transfer the wind and seismic loads covered by the scope of the IRC. Note that in the more severe wind and seismic regions addressed by the IRC, if the minimum connection is used, every fastener must be properly sized and placed. Note also that the often overlooked Footnotes f and g of IRC Table R602.3(1) require additional attachment for roof sheathing in areas with wind speeds of 100 mph or greater. The additional nailing provides enhanced diaphragm performance and prevents sheathing from pulling off in high wind events.

ROOF SHEATHING EDGE NAILING

Panel edge nailing

An important design consideration often overlooked in diaphragm connections is that the roof diaphragm actually ends at the supporting wall line and not at the end of the overhang. At the wall line, the panel "edge" nailing (6 inches on center) as specified in IRC Table R602.3(1) should be used to attach the roof sheathing.

FIGURE 1.11

Notice that the walls are still intact, just no longer in place. This was a panelized building and the wall panels were not attached sufficiently to the floor and roof to resist the wind loads. (Photo taken after Hurricane Andrew.)

FIGURE 1.12

This is an example of a garage door failure (on backside of structure) that led to pressurization of the garage wall. Insufficient attachment at the top and bottom caused the wall to blow out. This portion of the end wall is no longer available to provide bracing for the rest of the structure. (Photo taken after Hurricane Katrina.)

FIGURE 1.13

This figure shows insufficient attachment of the end walls of the structure. It also shows failures in the wall covering products used to weatherproof the structure. (Photo taken after a tornado in Evansville, Indiana.)

FIGURE 1.14

This building failed due to insufficient attachment of numerous structural elements, including receiving wall-to-roof, roof-to-wall, and wall-to-wall. The entire roof of this structure was lifted off in one piece and struck the house across the street. (Photo taken after Hurricane Iniki.)

For multi-story structures in high wind areas, builders may be accustomed to using strap-type anchors between floors. These greatly increase the structure's ability to stay together. While most often required for wind uplift, these straps also provide reinforcement to the lateral load path, if for no other reason than to prevent uplift forces from damaging the lateral load path.

5. Wall bracing

From a lateral load perspective, the walls support the roof and floor diaphragms through the use of bracing panels. The type, amount, and number of bracing panels are, of course, dependent on the magnitude of the lateral load. Stronger resistance (greater numbers of bracing panels) and reduced braced wall line spacing (interior braced wall lines) may be required in areas of high wind and/or seismic activity.

Failure of a braced wall line is evidenced by racking of the wall line. Racking occurs when a rectangular wall deforms to a parallelogram shape, in which the top and the bottom of the wall remain horizontal but the sides are no longer vertical. The purpose of wall bracing is to prevent such failures. Examples of racking in various degrees are shown in **FIGURE 1.15**.

FIGURE 1.15

Failures in wall bracing as indicated by wall racking

(File photos, both wind and seismic events.)

6. Wall-to-foundation connections

Just as the receiving walls must be attached to the foundation in order to resist the loads imposed, the bracing walls must be attached to the foundation. These connections are critical. Examples of inadequate connections between the walls and foundation are demonstrated in **FIGURES 1.16** and **1.17**.

FIGURE 1.16

Insufficient (or non-existent) anchorage of the walls to the foundation caused the whole structure to be pushed off the foundation in this wind event. Additional connection failures can be seen in the background. (Photo taken after a tornado in Southwestern Missouri.)

FIGURE 1.17

This picture is a close-up of insufficient foundation anchorage. Few bolts were used and washers were not used under nuts. Direction of bends indicated a shear failure of the bottom plate in a direction up and to the right hand corner of the picture. (Photo taken after a tornado in Southwestern Missouri.)

The 2009 IRC contains a number of new provisions that deal directly with the connection of braced wall lines to the structure above and below. These new provisions are an attempt to address a number of gaps in the load path that had existed in previous IRC editions. These provisions can be found in IRC Section R602.10.6 and will be discussed in greater detail in **CHAPTER 8**.

The solution

The previous photos may suggest that damage is probable in wind or seismic events but this is not always the case. In all of these examples, the damage to the building could have been prevented or minimized if the structure had been designed or built in accordance with the prevailing building code. In nearly all of these cases, there were similar structures in very close proximity that sustained little or no damage because they were built to the requirements of the building codes that existed when they were constructed.

This book explains in detail how to use the IRC to provide adequate wall bracing so that a house can be built to resist lateral loads. **FIGURES 1.18** and **1.19** show two examples of how properly built homes can survive even the most severe wind or seismic events.

FIGURE 1.18

After the 1964 Good Friday earthquake in Alaska, a house built with continuously sheathed plywood walls retained its box-like structure, even after a significant portion of the ground below shifted more than 20 feet. Note that the walls are strong enough to cantilever the structure over the collapsed foundation. The house to the left appears to have performed well without any noticeable damage. (Photo taken after the Great Alaska Earthquake.)

FIGURE 1.19

After Hurricane Andrew struck in 1992, a waterfront house in Florida stands in near perfect condition, even after being battered by significant winds and water. Such cases of survival demonstrate that a proper load path works. (Photo taken after Hurricane Andrew.)

CHAPTER 2

Wall Bracing – How it Works and Code History

As discussed in the preceding chapter, wind and seismic events subject structures to loads acting parallel to the ground (lateral loads). Regardless of how the structure is designed, whether through the use of prescriptive braced walls or engineered shear walls, elements must be built into the structure to resist these lateral loads. Without either of these elements, there is a real potential for structural failure. Both braced walls and shear walls serve the same purpose in the lateral load path.

Braced walls (prescribed) vs. shear walls (engineered)

Bracing provisions in the model codes are prescribed for conventional construction. From the user's perspective, there is no "engineering" required in conventional construction since the construction requirements are specified in

PRESCRIPTIVE VS. ENGINEERED

Prescriptive Construction	Engineered Design
Limitations:	Applications:
• 3 stories max.	• Any size/shape within IBC limits
• Wind < 110 mph (< 100 mph in hurricane-prone regions)	• Wind – no limit
• SDC A-D$_2$	• Seismic – no limit
• Many others (see IRC Chapter 3)	• Governed by engineers' calculations
Typically uses braced wall panels without hold downs to prevent walls from racking.	Typically uses shear walls with wood structural panels and pre-engineered hold downs (other than anchor bolts) to prevent walls from racking and overturning.

the code. For structures, or portions of structures, that do not meet conventional construction parameters, engineered design per the International Building Code (IBC) or other referenced standards is required. The engineered counterparts to braced walls are shear walls. Both provide racking resistance to lateral loads, but each comes from a distinctly different set of model-code provisions.

What's the difference between a braced wall panel and shear wall?

Braced wall panels, the primary focus of this book, come from prescriptive building codes. In the IRC, wall bracing is prescribed – required bracing lengths are provided in IRC tables – and few, if any, calculations are necessary.

Shear walls, on the other hand, are designed by a design professional and have specific design values depending on fastener spacing, fastener size, sheathing thickness and framing species. They usually require manufactured hold downs to resist being overturned. The shear wall must resist the loads that are calculated through engineering analysis. Shear walls are associated with the design provisions of the IBC.

What is bracing and how does it work?

The framing elements of a typical stud wall are shown in **FIGURE 2.1**. When subjected to lateral loads, the stud wall without bracing has very little racking resistance because each of the joints act as a hinge (shown as a small circle in **FIGURE 2.1**). It takes very little lateral load to rack the wall, turning it into a parallelogram.

FIGURE 2.1

Bare stud wall has no lateral load resisting capacity

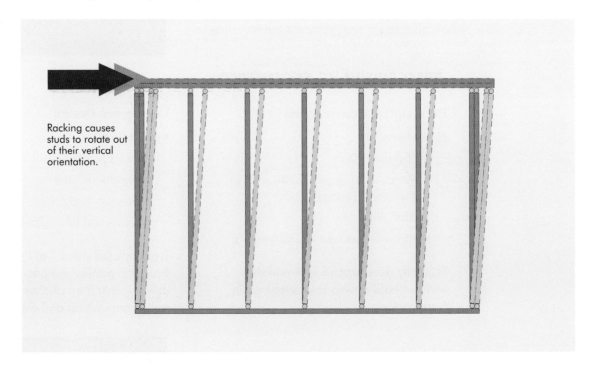

Racking causes studs to rotate out of their vertical orientation.

In prescriptive (also referred to as conventional) construction, there are two common ways to provide a stud wall with lateral load resistance. One method is to use a diagonal 1x4 let-in brace between 45 and 60 degrees to the horizontal, as shown in **FIGURE 2.2**. The other method is through the attachment of relatively rigid rectangular panels (**FIGURE 2.3**) to the stud wall, as shown in **FIGURE 2.4**. (There is a third method permitted – the use of diagonal wood boards along the entire wall surface – that is seldom used today and, therefore, is not addressed in this chapter.)

FIGURE 2.2

Stud wall with let-in brace

The diagonal brace and two nails into the top and bottom plates and each stud it crosses prevent the wall from racking

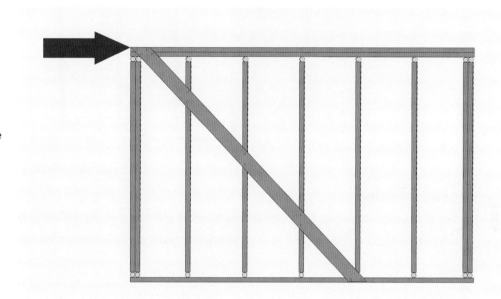

FIGURE 2.3

Rectangular panel products

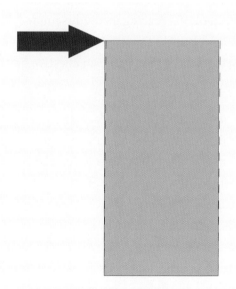

Rectangular panel products (and portland cement lath and plaster) resist being forced out of shape by their physical properties and nail attachment to studs.

FIGURE 2.4

Stud wall with panel bracing

Perimeter nailing prevents the wall from racking. (Intermediate nailing not shown for clarity.)

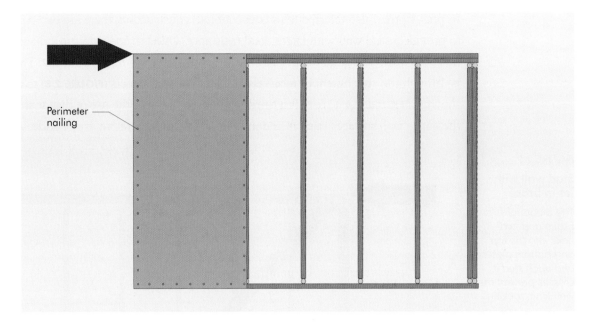

Perimeter nailing

Let-in bracing

Let-in bracing, known in the 2009 IRC as Method LIB bracing and shown in *FIGURE 2.2*, is the original method of bracing that was used before panel-type sheathing products were developed.

The effectiveness of let-in bracing depends on the craftsmanship of the framer when cutting the notches for the 1x4 brace, the nails that attach the brace to the top and bottom plates, and the condition and species of the brace. Since there are only two nails at each point where the 1x4 brace crosses the framing, the strength of this bracing method is limited. Up until the last 50 years or so, residential structures were relatively small and divided into many small rooms. For these types of houses, let-in bracing was (and still may be) an effective bracing method.

A number of manufacturers have developed proprietary metal straps designed to function like a 1x4 brace. These straps either require no modification of the studs or just a shallow saw kerf to inset the strap. Generally, the thin flat straps only resist tension and have to be used in pairs to make an "X" pattern, while the straps with an "L" or "T" cross section resist tension and compression and normally can be used singly, like 1x4 let-in bracing. These proprietary products are recognized for use in code-conforming construction in ICC Evaluation Service (ICC-ES) or other reports. Copies of these reports, which include use restrictions, are available at www.icc-es.org or other agencies' websites. It is important to review the details of all relevant reports, as well as the manufacturer's installation recommendations carefully because some of these strap-type anchors are approved for use as bracing and some are meant only for temporary bracing during construction.

Let-in bracing and proprietary strap-type products have limited structural capacity. As such, the IRC and proprietary ICC-ES reports limit their use to areas of low wind and seismic loads.

Panel-type bracing (and portland cement lath and plaster)

Offering considerably more structural capacity than let-in bracing, and consequently of greater use, are panel-type products. While installation and fastening requirements vary greatly, all panel-type bracing methods work in the same manner and have the same capacity for wind resistance. In regards to seismic resistance, Method WSP (wood structural panel) provides greater strength and is rated at the upper end of the bracing capacity spectrum, while Method GB (gypsum board) is rated at the less effective end. The panel bracing methods have the following properties:

- The panel-type products are manufactured in a rectangular shape and, through their physical properties, resist being forced out of this shape. See *FIGURE 2.3*.

- Fasteners are used to attach panel products to wall framing members. The spacing of the fasteners around the perimeter of the panel typically ranges from 3 to 7 inches on center depending on the bracing method used. The spacing of the fasteners in the middle of the panel (commonly called the "field") typically ranges from 6 to 12 inches depending on the bracing method selected.

- When panel products are attached to a stud wall, such as the one shown in *FIGURE 2.1*, they provide racking resistance to the whole wall, as shown in *FIGURE 2.4*.

- Portland cement lath and plaster is considered panel bracing because, although it is formed in place, it performs in a similar manner and capacity as the panel bracing methods.

- Bracing Method GB (gypsum board) is similar to the other panel-type bracing methods except that its capacity is about half as strong as other panel products when applied to one side of the wall.

The ability of a panel to resist loads depends on the physical properties of the panel, primarily its strength and rigidity. The strength of the panel is imparted to the wall by the fasteners installed at its perimeter to connect it to the wall framing. The number, size and placement of the nails (or other fasteners), along with the structural capacity of the framing and physical properties of the panel (such as fastener bearing strength, tear resistance and fastener pull-through resistance), affect the attachment capacity. Variations in bracing panel physical properties and prescribed attachments account for the different installation requirements for the various panel type bracing methods. These differences are discussed in more detail in *CHAPTER 5*.

History of wall bracing

TABLE 2.1

Summary of wall bracing history in model codes

Year	Bracing Provisions
1920s	"...thoroughly and effectively angle braced..."
1950s	"...thoroughly and effectively angle braced or sheathed..."
1970s	"...effectively and thoroughly braced at each end and...every 25 feet... by one of the following methods..."
	"...braced panel must be at least 48 inches in width..."
1994	ADD: 32-inch alternate braced panel = 48-inch braced panel
2000	ADD: Minimum bracing percentages (based on load and story location)
	ADD: Continuous sheathing method
2003	ADD: Braced wall line maximum spacing
2006	ADD: Expanded alternate braced panels
	ADD: Portal frames
2009	ADD: Additional narrow-width bracing methods
	ADD: Wind bracing table based on engineering
	ADD: Changed method numbers to abbreviations
	ADD: Separate wind and seismic bracing tables

Wall bracing requirements are not new to the codes. In fact, several of the current bracing methods, including let-in bracing, diagonal wood boards and portland cement plaster, reflect conventional construction practices that were common 50 years ago. While the Uniform Building Code's (UBC) history of bracing is presented in brief below, note that it is virtually identical to the history of the Standard Building Code (SBC) and the Basic National Building Code (BNBC). Bracing provisions in all three of the legacy model codes responded to the evolution of the bracing provisions in a similar manner. A summary of the changes over the years to the model codes' wall bracing provisions is provided in *TABLE 2.1*.

Wall bracing requirements date back to the first edition of the UBC in 1927. For Type V construction, which would include the types of residential construction seen today, the 1927 UBC required:

> ***Section 2205 (4th paragraph).*** *All exterior walls and partitions shall be thoroughly and effectively angle braced.*

Similar requirements were in Section 2507(e) for wood-frame construction.

In the 1920s, panel sheathing products were not used. Thus, the means of bracing was presumably let-in bracing, or some other type of angle bracing.

The UBC bracing provisions remained unchanged until 1952, which was also the year that vertical and horizontal diaphragm tables appeared. The text of the 1952 UBC is virtually identical to that for Type IV construction. In the 1952 UBC, however, plywood panels were added by virtue of being an "approved panel."

> **2521(f) –** …*All exterior walls and partitions shall be thoroughly and effectively angle braced or sheathed with approved panels adequately nailed along all edges.*

These provisions were unchanged until 1970. The 1970 UBC added several pages to describe conventional construction provisions. These sections included expanded descriptions of different bracing methods and prescriptive provisions, such as the distance between bracing panels. The text of the 1970 UBC bracing provisions in Section 2518 was:

> **5. Bracing.** *All exterior walls and main cross stud partitions shall be effectively and thoroughly braced at each end, or as near thereto as possible, and at least every 25 feet of length by one of the following methods:*
>
> **a.** *Nominal 1-inch by 4-inch continuous diagonal braces let into studs…*
> (nearly identical to today's Method LIB (let-in) bracing)
>
> **b.** *Wood boards of 5/8-inch net minimum thickness applied diagonally on studs…*
> (nearly identical to today's Method DWB (diagonal wood boards))
>
> **c.** *Plywood sheathing…* (nearly identical to today's Method WSP (wood structural panel) bracing)
>
> **d.** *Fiberboard sheathing…* (nearly identical to today's Method SFB (structural fiberboard sheathing) bracing)
>
> **e.** *Gypsum sheathing…* (nearly identical to today's Method GB (gypsum board) bracing)
>
> **f.** *Particleboard sheathing…* (nearly identical to today's Method PBS (particleboard sheathing) bracing)
>
> *For methods b, c, d, e and f, the braced panel must be at least 48 inches in width…*

In the 1994 UBC, a 32-inch alternate braced wall panel was added that prescribed sill plate anchorage, foundation details, and hold-down devices. This 32-inch alternate braced wall panel with hold downs is still in today's code. The alternate braced wall panel was essentially a prescriptive shear wall that was designed to be equivalent to a Method WSP (wood structural panel) wall segment and thus is a suitable substitution for all bracing methods.

Between the 1970 and the 1997 UBC, *"effectively and thoroughly braced"* was more clearly defined.

Other model codes, such as the Council of American Building Officials (CABO) One- and Two-Family Dwelling Code (O&TFDC), the International Code Council (ICC) International One- and Two-Family Building Code, the Southern Building Code Congress International (SBCCI) Standard Building Code, and the Building Officials and Code Administrators International (BOCA) National Building Code, also had similar bracing provisions and evolution.

During the development of the 2000 IRC, the IRC Code Development Committee selected and combined code provisions from the existing model codes to form the International Residential Code.

Some of the new, notable wall bracing provisions in the 2000 IRC were:

- Wall bracing was required to meet a minimum percentage of the braced wall line. The amount depended on the load (wind speed or Seismic Design Category) and how many stories were above the braced wall line.

- A continuous wood structural panel bracing method was added, permitting wall bracing elements as narrow as 24 inches to be used to meet the percent bracing requirement.

In the 2003 IRC, braced wall line spacing limits were added. Also, an option for continuous wood structural panel sheathing (based on the perforated shear wall method defined in the IBC) was available to allow narrower braced wall panels. This method required the braced wall lines to be fully sheathed with wood structural panels.

In the 2006 IRC, the 32-inch-wide alternates were changed to permit wall bracing elements as narrow as 28 inches for certain cases. A portal frame with hold downs – a narrow vertical element that is attached at the plate level below and to the header above in a way that will allow it to act as a bracing unit – was also added, which permitted bracing elements as narrow as 16 inches in certain cases. In addition, the portal frame was permitted to be used without hold downs for bracing next to garage doors (in Seismic Design Categories (SDC) A, B and C, for use with up to one story above) when used in conjunction with continuously sheathed wood structural panel walls (IRC Table R602.10.5, Footnote c).

In the 2009 IRC, many changes were made to the wood-frame bracing provisions:

- Most significant was the development of wind bracing tables based on engineering principles.

- Separate tables for wind and seismic bracing were developed.

- The IRC was reorganized to consolidate all of the bracing provisions for wood-frame construction into the Chapter 6 bracing section.

- New bracing methods were added to increase the choices available to the builder and to reflect ongoing product research.

- Bracing methods were defined by abbreviations instead of method numbers. For example, wood structural panel bracing, formerly referred to as Method 3, became Method WSP. Gypsum board bracing became Method GB, let-in bracing became Method LIB, structural fiberboard sheathing became Method SFB, etc.

- The number of narrow wall bracing alternates grew from two to five.

- Method SFB (structural fiberboard sheathing) was recognized as a continuous sheathing method for use in areas of low wind and earthquake loads.

The bracing provisions of the 2009 IRC have more than doubled in page count over the 2006 bracing provisions, due to additional bracing options and new figures that make the provisions easier to understand and apply. Better understanding of the provisions leads to better application and enforcement.

In short, model codes have required wall bracing for more than 70 years. Like many other code requirements, the changes to wall bracing provisions have accelerated in recent years.

Why are bracing provisions changing?

The simple answer is that we don't build houses the way we used to!

In the 1960s, when more prescriptive bracing provisions began appearing in model building codes, houses were much smaller (seldom over two stories) and had a greater number of interior walls and fewer windows. See **FIGURE 2.5**.

FIGURE 2.5

Typical single-family residence built in the 1960s.

Houses today are on average twice as large as those built in the 1960s. Popular features include walk-out basements or third stories, two-story entrance foyers and stairways, great rooms, window walls, two-car garages, complex roof lines and dramatic stairways that leave large openings in the floor diaphragms. Any one of these features can have a negative impact on the structural performance of a building if not accounted for in the structural frame. **FIGURE 2.6** is an example of a modern house containing many of these features.

FIGURE 2.6

A residence more typical of the type being built today.

Another factor is the prevalence of new building materials, including nonstructural sheathing products. In the 1960s, the building industry had little concern about wall bracing because the majority of sheathing products available for use were structural to one degree or another. Today, there are a number of nonstructural sheathing products that are popular for energy conservation. Builders are compelled by market pressures to balance the structural and energy requirements of the building. As a result, builders often inadvertently minimize the structural sheathing in an effort to maximize the insulating materials. The trends toward energy efficiency, new construction methods and materials, and other changes in modern residential design are driving the seemingly endless changes to the prescriptive bracing requirements.

Many of the changes to wall bracing in the last 15 years have been made in attempt to permit narrower bracing segments, as space for multiple 4-foot bracing segments is not always available in today's home designs. Garage door openings are the classic example. Seldom do garage door openings have the full 48 inches of bracing length necessary on either side of the opening, as required by the codes for over 20 years. But narrow bracing options now make it easier for home designers to meet prescriptive code requirements while enhancing the architectural appearance of the home.

TODAY'S COMPLEX HOUSE

For these reasons, along with high gravity or lateral loads, many homes constructed in certain regions are engineered. For example, in Tacoma, Washington, 90 percent of all new houses constructed have an engineered load path. This is also true for new homes in parts of Florida, California, Oregon and other states.

The one thing that has not changed, however, is the need to ensure that the new housing stock is safe. Safety concerns prompted the evolution of building codes in the 1950s and 1960s, and is expected to continue into the future. Today's houses are very different from those built 50 years ago, and the houses of the future will likely differ significantly from today's home designs.

CHAPTER 3

Getting Started

So far, we have addressed why bracing is necessary and how it works to resist lateral loads. Now we will begin applying this knowledge specifically to the bracing provisions of the IRC.

Information on construction documents

New to the 2009 IRC is a provision that requires wall bracing information to be included on construction documents. The purpose of this provision is to place the responsibility for detailing the wall bracing on the building designer rather than the building official or builder. This is addressed in this excerpt from the 2009 IRC:

> **R106.1.1 Information on construction documents.** *Construction documents shall be drawn upon suitable material. Electronic media documents are permitted to be submitted when approved by the building official. Construction documents shall be of sufficient clarity to indicate the location, nature and extent of the work proposed and show in detail that it will conform to the provisions of this code and relevant laws, ordinances, rules and regulations, as determined by the building official. Where required by the building official, all braced wall lines, shall be identified on the construction documents and all pertinent information including, but not limited to bracing methods, location and length of braced wall panels, foundation requirements of braced wall panels at top and bottom shall be provided.*

Bracing terminology

When describing wall bracing, the IRC uses the terms *braced wall panel, braced wall panel spacing, braced wall line* and *braced wall line spacing*. The user must understand what each of these terms means in order to apply the IRC bracing requirements. **FIGURE 3.1** shows how these terms relate to an actual structure.

FIGURE 3.1

Braced wall panels, braced wall panel spacing, braced wall lines and braced wall line spacing

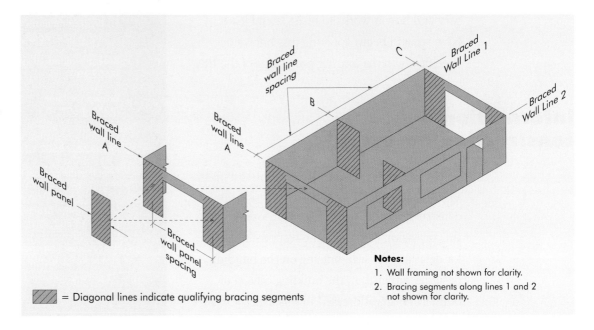

= Diagonal lines indicate qualifying bracing segments

Notes:
1. Wall framing not shown for clarity.
2. Bracing segments along lines 1 and 2 not shown for clarity.

Braced wall panel

From Section R202 of the 2009 IRC:

> **Braced wall panel.** *A full-height section of wall constructed to resist in-plane shear loads through interaction of framing members, sheathing material, and anchors. The panel's length meets the requirements of its particular bracing method and contributes toward the total amount of bracing required along its braced wall line in accordance with Section R602.10.1.*

Put in simpler terms, braced wall panel describes a code-qualified bracing element. The name is probably derived from the fact that most of the recognized methods of bracing use panel-type products. Even let-in bracing is often referred to as a bracing panel. "Panel" is somewhat of a misnomer since it actually describes a wall "section," "segment," or "unit." As such, the terms *braced wall section, braced wall segment* and *bracing unit* are often used interchangeably.

Each braced wall panel must extend the full height of the wall – from the bottom plate to the top of the double top plates. A "panel" may be constructed from more than one piece of sheathing. For example, a 6-foot long braced wall panel may be constructed by joining a 4-foot long panel with a 2-foot long panel.

Braced wall panels have a height and a length dimension. The permitted height of a braced wall panel ranges from 8 to 12 feet. In some cases, the panel height is limited to 10 feet. The length dimension is measured parallel to the length of the wall. For example, the length of a 4x8 oriented strand board (OSB) bracing panel placed with the 8-foot dimension in the up-and-down direction is 4 feet. Knowing the length of a braced wall panel is important because the various bracing methods have different required minimum lengths. Also, the combined length of individual braced wall panels must total or exceed a required minimum length for the braced wall line.

What is the required minimum length of a braced wall panel? It varies from 16 inches (Method PFH (intermittent portal frame) or Method CS-PF (continuous portal frame)) in an 8-foot wall to 12 feet (Method LIB (let-in bracing)) at a 45 degree slope in a 12-foot high wall. The length depends on the bracing method used and the height of the wall. In addition to a number of specific narrow-width braced wall panels, in the 2009 IRC, traditional 4-foot braced wall panels are permitted to be used in lengths of less than 4 feet, provided certain reductions are taken into account when computing the amount of bracing necessary for the braced wall line. (See IRC Table R602.10.3, reproduced as **TABLE 5.5** in **CHAPTER 5**.)

Bracing methods

The bracing methods are detailed in **CHAPTER 5**.

The 2009 IRC differs from earlier editions in the way it identifies bracing methods. The term *intermittent* is used to identify bracing methods that can be placed in discrete locations along a braced wall line, while the term *continuous* is used to identify bracing methods that require the whole wall line to be sheathed (Method CS-WSP (continuous wood structural panel sheathing) or Method CS-SFB (continuous structural fiberboard sheathing)).

There are eleven different intermittent bracing methods included in IRC Section R602.10.2 of the 2009 IRC, each with its own minimum length and installation requirements.

- Eight of the eleven intermittent methods are Methods 1-8 from previous editions of the IRC, two methods are the alternate methods permitted in the 2006 IRC, and there is one newly added narrow-width method for use next to garage doors in SDC A, B and C.

- The two alternate methods brought forward from the 2006 IRC differ in that one is a portal frame (Method PFH (intermittent portal frame)) and the other is not (Method ABW (alternate braced wall)).

- The newly added narrow-width method in the 2009 IRC is called the Method PFG (intermittent portal frame at garage door openings in Seismic Design Categories A, B and C).

In the 2009 IRC, the last three methods are listed with the original 8 methods because they can also be used intermittently along a braced wall line.

The 2009 IRC also recognizes four continuous bracing methods that require continuous sheathing on all wall areas of the wall line. As walls with continuous sheathing are generally considered to be stiffer and stronger than similar walls with intermittent bracing, the continuous methods allow for bracing solutions that are considerably narrower than the widths required for the intermittent methods.

- Three of these bracing methods are for use with continuous wood structural panel sheathing (IRC Section R602.10.4). These methods are detailed in **CHAPTER 5**.

- The fourth continuous method, Method CS-SFB (continuous structural fiberboard sheathing, IRC Section R602.10.5), is a new addition and detailed in **CHAPTER 5**. The provisions for this bracing method are formatted differently than the other continuous sheathing methods due to multiple factors, including the text of the original code change, the reformatting of the section due to other code changes, and the ICC staff's limited ability by regulation to editorialize as various code changes are blended. This will undoubtedly be addressed in future editions of the IRC.

- As detailed in **TABLE 3.1**, the traditional method of referring to bracing methods by number has been replaced in the 2009 IRC with abbreviations. The abbreviations for the continuous sheathing methods each start with CS (e.g., Method CS-PF is the continuous portal frame bracing method).

Although in previous editions of the IRC the terms "length" and "width" were often interchanged, the 2009 IRC uses the term "length" more consistently throughout the bracing provisions.

TABLE 3.1

2009 IRC Bracing Methods

Intermittent Bracing Methods

2006 IRC Designation	2009 IRC Designation	2009 IRC Section	Construction Description
Method 1	LIB		Nominal 1 x 4 let-in bracing or approved metal strap
Method 2	DWB		Wood boards, 5/8": nominal thickness applied diagonally
Method 3	WSP		Wood structural panel sheathing
Method 4	SFB	R602.10.2	Structural fiberboard sheathing
Method 5	GB		Gypsum sheathing/board
Method 6	PBS		Particleboard sheathing
Method 7	PCP		Portland cement plaster
Method 8	HPS		Hardboard panel siding
Alternate braced wall panels	ABW	R602.10.3.2	Bracing panel with hold downs. See IRC Table R602.10.3.2 and IRC Figure R602.10.3.2.
Alternate braced wall panel adjacent to a door or window opening	PFH	R602.10.3.3	Portal frame with hold downs. See IRC Figure R602.10.3.3.
New for IRC 2009	PFG	R602.10.3.4	Portal frame without hold downs at garage door for SDC A, B and C. See IRC Figure R602.10.3.4.

Continuous Sheathing Bracing Methods
Using Wood Structural Panel Sheathing

2006 IRC Designation	2009 IRC Designation	2009 IRC Section	Construction Description
Continuous wood structural panel sheathing	CS-WSP	R602.10.4 and Table R602.10.4.2	Continuously sheathed wood structural panel
No designation (Table R602.10.5, Footnote b)	CS-G	R602.10.4 and Table R602.10.4.2	4:1 aspect ratio wall segment adjacent to garage door
No designation (Table R602.10.5, Footnote c)	CS-PF	R602.10.4.1.1 and Table R602.10.4.2	Portal frame without hold downs. See IRC Figure R602.10.4.1.1.

Continuous Sheathing Bracing Method
Using Structural Fiberboard Sheathing

2006 IRC Designation	2009 IRC Designation	2009 IRC Section	Construction Description
New for IRC 2009	CS-SFB	R602.10.5	Continuously sheathed structural fiberboard sheathing

Braced wall panel spacing

IRC Section R602.10.1.4 requires bracing panels to be spaced not farther than 25 feet on center. The spacing is measured from the centerline of the bracing panel to the centerline of adjacent bracing panels in the same wall line. See **FIGURES 3.2** and **3.3**.

Note: Measuring between 4-foot bracing panels that are a part of wider bracing segments, as shown in **FIGURE 3.3**, is a correct interpretation of the code. Because the intent of the code is to limit the unbraced portions of the wall to 21 feet – the clear distance between braced wall panels – the solution provided in **FIGURE 3.3** is acceptable.

FIGURE 3.2

Braced wall panel spacing

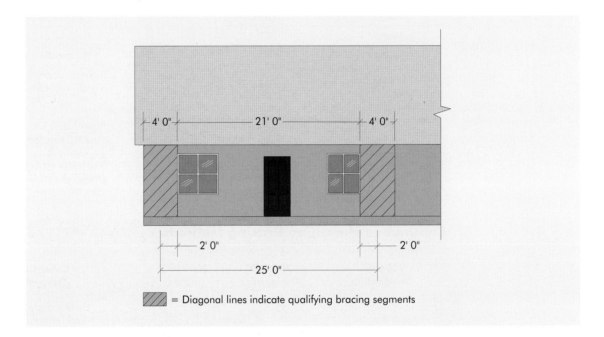

= Diagonal lines indicate qualifying bracing segments

FIGURE 3.3

Braced wall panel spacing when braced wall sections are more than 4 feet long

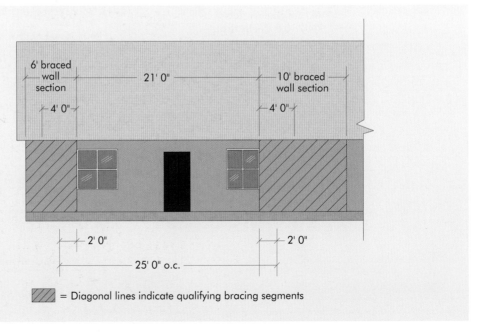

= Diagonal lines indicate qualifying bracing segments

Braced wall line

From Section R202 of the 2009 IRC:

> **"Braced wall line.** *A straight line through the building plan that represents the location of the lateral resistance provided by the wall bracing."*

FIGURE 3.1 shows two braced wall lines (1 and 2) in the longitudinal direction and three braced wall lines (A, B and C) in the transverse direction. Each wall line is made up of a number of braced wall panels. Not every braced wall line has to be continuous, as shown in braced wall line B. Furthermore, the code permits offsets of up to 4 feet (IRC Section R602.10.1.4). If sections of a wall line are offset by 4 feet or less, they are assumed to act in combination to resist lateral load. If the sections are farther apart than 4 feet, they must be considered as separate braced wall lines. Offsets in braced wall lines are further discussed in **CHAPTERS 7** and **10**.

Braced wall lines are not always exterior walls, as braced wall line B in **FIGURE 3.1** illustrates. An interior braced wall line may be required – depending on the size of the house, the wind speed, or Seismic Design Category (SDC) – to supplement the exterior braced wall lines. Interior braced wall lines have requirements similar to exterior wall lines in terms of bracing length, panel location, wall line offsets and attachments. The foundation requirements and other attachment requirements are covered in **CHAPTER 8**.

In the 2006 IRC, a major challenge in working with braced wall lines was that if a building was not a simple rectangle, it was difficult to determine the beginning and end points of braced wall lines, which made measuring the line length problematic. See *Length of the braced wall line* on page 35 for more information.

Continuously sheathed braced wall line

The definition is new in Section R202 of the 2009 IRC:

> **Braced wall line, continuously sheathed.** *A braced wall line with structural sheathing applied to all sheathable surfaces including the areas above and below openings.*

In the 2006 IRC, the provisions required all walls on all stories to be continuously sheathed. The intent of the provisions was clarified in the 2007 Supplement and brought forward into the 2009 IRC: continuously sheathed braced wall lines were to apply separately to each wall line. For example, just the front side of an attached garage could be continuously sheathed to permit narrow walls on either side of the garage door, while the remaining walls of the house could use other IRC-approved bracing methods. Continuously sheathed braced wall lines are covered in Sections R602.10.4 and R602.10.5 of the 2009 IRC and discussed in **CHAPTER 5**.

Effective (imaginary) braced wall lines

The concept of the effective – or "imaginary" – braced wall line is based on the principle that when braced wall panels are located no more than 4 feet from a designated braced wall line, it is not necessary to have a bracing panel on the designated braced wall line. The designated braced wall line does not have to coincide with the wall line section at all. A line of lateral resistance is known to exist between panels grouped together in close proximity, as shown on the left side of **FIGURE 3.4**. The location of the designated effective (imaginary) wall line is used for measuring the braced wall line spacing required for use of IRC Tables R602.10.1.2(1) and R602.10.1.2(2). All the wall sections that make up the effective (imaginary) braced wall line must meet the same requirements (connection, anchorage, etc.) as a regular braced wall line. In addition, the lengths of the braced wall panels on the various wall sections added together must be equal to or greater than the length of bracing required by IRC Tables R602.10.1.2(1) and R602.10.1.2(2).

Note that the provisions for effective (imaginary) braced wall lines are not limited to exterior walls. In the 2009 IRC, there are few differences between the requirements for interior and exterior braced wall lines; in fact, the term "interior braced wall line" was eliminated from the 2009 edition specifically to prevent variations in how interior and exterior braced wall lines are used.

While not given by the name "effective (imaginary) braced wall line," this concept is addressed in IRC Section R602.10.1.4 and Figure R602.10.1.4(4), which is reproduced in **FIGURE 3.4**. In this figure, the effective (imaginary) braced wall line is presented as "BWL A".

When the concept was debated by the ICC Ad Hoc Bracing Committee, there was concern that this method would be used to reduce the distance between braced wall lines by arbitrarily moving the effective (imaginary) braced wall line inward from one or both exterior wall lines up to 4 feet. This is not the intent of the provision. The provision was placed in the code to permit a closely spaced grouping of wall segments to be counted as a single braced wall line to accommodate popular architectural variations in exterior wall lines. It is not appropriate to use effective (imaginary) braced wall lines as a tool to reduce the distance between braced wall lines.

FIGURE 3.4

Effective (imaginary) braced wall lines

IRC Figure R602.10.1.4(4) Braced wall line spacing

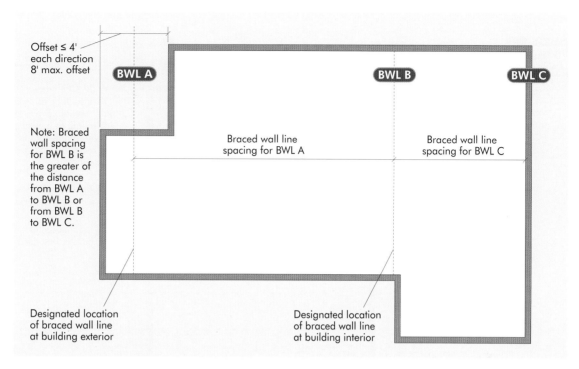

Offset ≤ 4' each direction 8' max. offset

BWL A   BWL B   BWL C

Note: Braced wall spacing for BWL B is the greater of the distance from BWL A to BWL B or from BWL B to BWL C.

Braced wall line spacing for BWL A

Braced wall line spacing for BWL C

Designated location of braced wall line at building exterior

Designated location of braced wall line at building interior

Length of the braced wall line

The length of the braced wall line is defined in IRC Section R602.10.1 as follows:

> *"The length of a braced wall line shall be measured as the distance between the ends of the wall line."*

While this definition may seem simplistic, its significance becomes apparent as house plans deviate from simple rectangles. With permitted braced wall line offsets, enclosed porches, wall bump outs, etc., it can be difficult for all parties to agree on where the beginning and end points for a given braced wall line are. This can become increasingly confusing when incorporating the new effective (imaginary) braced wall lines, as permitted in IRC Figure R602.10.1.4(4). Because of the effective (imaginary) braced wall line concept, the following provisions were added in IRC Section R602.10.1:

> **R602.10.1 Braced wall lines.** *The end of a braced wall line shall be considered to be either:*
>
> - *The intersection with perpendicular exterior walls or projection thereof.*
>
> - *The intersection with perpendicular braced wall lines.*
>
> - *The end of the braced wall line shall be chosen such that the maximum length results.*

FIGURE 3.5a

Actual wall lines/projection

IRC Section R602.10.1

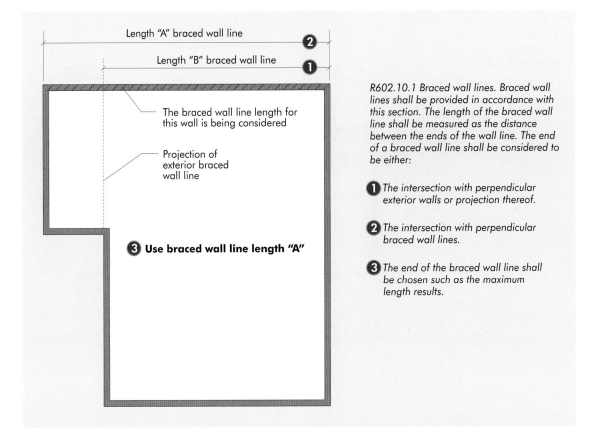

Length "A" braced wall line ❷

Length "B" braced wall line ❶

The braced wall line length for this wall is being considered

Projection of exterior braced wall line

❸ **Use braced wall line length "A"**

R602.10.1 Braced wall lines. Braced wall lines shall be provided in accordance with this section. The length of the braced wall line shall be measured as the distance between the ends of the wall line. The end of a braced wall line shall be considered to be either:

❶ The intersection with perpendicular exterior walls or projection thereof.

❷ The intersection with perpendicular braced wall lines.

❸ The end of the braced wall line shall be chosen such as the maximum length results.

FIGURE 3.5b

Effective (imaginary) braced wall lines

IRC Section R602.10.1

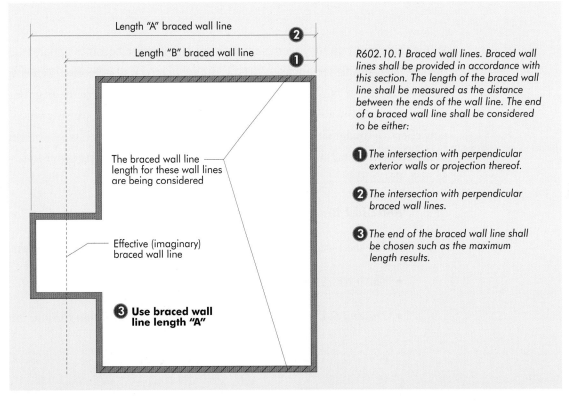

Length "A" braced wall line ❷

Length "B" braced wall line ❶

The braced wall line length for these wall lines are being considered

Effective (imaginary) braced wall line

❸ **Use braced wall line length "A"**

R602.10.1 Braced wall lines. Braced wall lines shall be provided in accordance with this section. The length of the braced wall line shall be measured as the distance between the ends of the wall line. The end of a braced wall line shall be considered to be either:

❶ The intersection with perpendicular exterior walls or projection thereof.

❷ The intersection with perpendicular braced wall lines.

❸ The end of the braced wall line shall be chosen such as the maximum length results.

These provisions are meant to address the following two conditions, which can arise from the addition of an effective (imaginary) braced wall line:

1. When the perpendicular braced wall line is on the exterior wall or inside of the perpendicular wall. See *FIGURE 3.5a*.

2. When the perpendicular braced wall line is located outside (away) from the location of the perpendicular line. See *FIGURE 3.5b*.

Braced wall line spacing

As shown in *FIGURE 3.1*, braced wall line spacing is the distance between parallel braced wall lines. In preceding editions of the IRC, braced wall line spacing was fixed at 25 or 35 feet, based on the Seismic Design Category (SDC). There were, however, provisions for increasing wall line spacing. In the 2009 IRC, the length required for wind bracing is based on the braced wall line spacing, but spacing for seismic wall bracing is again fixed at 25 foot intervals with adjustments permitted. In previous IRC editions, there was a single table for both wind and seismic conditions. The new IRC replaces the single table with two tables: one for wind loads and one for seismic loads. Each table addresses braced wall line spacing independently.

The 2009 IRC eliminates the need for a fixed braced wall line spacing limitation in areas of the country where the amount of bracing required for wind exceeds the amount of bracing required for seismic. The new wind bracing table (IRC Table R602.10.1.2(1)) provides the required amount of bracing as a function of braced wall line spacing. The table recognizes braced wall line spacings up to 60 feet.

The seismic bracing table (IRC Table R602.10.1.2(2)) is presented in a slightly different manner. Braced wall line spacing is limited to 25 feet (increasable to 35 feet). The table provides the length of bracing required based on the length of the braced wall line. This is reasonable, as the force on the structure is proportional to the mass (length) of the building in-line with the seismic force. This will be discussed further in *CHAPTER 5*.

An example for calculating the required increase in braced wall line length based on a braced wall line spacing greater than 25 feet is provided in *FIGURE 3.6*.

FIGURE 3.6

Example for calculating the increase in braced wall line length for wall line spacing greater than 25 feet

Per IRC Table R602.10.1.2(3) (SDC A, B and C) and Section R602.10.1.5 (IRC Table R602.10.1.5)

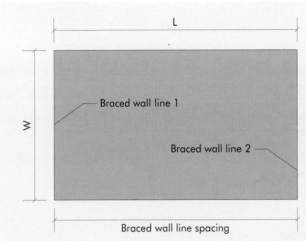

The braced wall line spacing in IRC Table R602.10.1.2(2) is limited to 25 feet or less, however:

For townhouses in SDC C – No adjustment required for spacing up to 35 feet, bracing length multiplied by 1.43 for over 35 feet to 50 feet (IRC Table R602.10.1.2(3)). Note that IRC Section R301.2.2 exempts all residential buildings in SDC A and B from the seismic provisions of the code.

For all structures in SDC D_0, D_1, D_2 – No adjustment required for spacing up to 35 feet to accommodate a single room not to exceed 900 square feet in each dwelling unit, otherwise braced wall line lengths should be adjusted in accordance with IRC Table R602.10.1.5, with certain limitations. Braced wall line spacing greater than 35 feet is not permitted.

Example: For townhouses in SDC C – A townhouse is designed with 42 feet between braced wall lines 1 and 2. What is the braced wall line length adjustment required for this braced wall line spacing? The distance between braced wall lines is between 35 and 50 feet; therefore, the braced wall line length must be multiplied by 1.43 (IRC Table R602.10.1.2(3)). The adjustment value may be interpolated.

For all structures in SDC D_0, D_1 and D_2 – Can the same braced wall line spacing of 42 feet be used? No, the exception to IRC Section R602.10.1.5 only permits braced wall line spacing to be increased to 35 feet, and then only after adjustment and with additional limitations on diaphragm aspect ratio and top plate splice requirements.

IRC Table R602.10.1.2(3) permits townhouses in SDC C to have a braced wall line spacing of up to 35 feet without penalty (a multiplier of 1.0 is provided) and provides a multiplier for spacings up to 50 feet. Although IRC Table R602.10.1.2(3) refers to townhouses in SDC A, B and C, note that all structures in SDC A and B are exempt from the seismic requirements of the code (IRC Section R301.2.2). Since this table refers to seismic bracing, with respect to townhouses in SDC A, B and C, it actually only refers to townhouses in SDC C. For townhouses in SDC D_0 and higher, refer to the braced wall line spacing provisions in IRC Table R602.10.1.5.

IRC Table R602.10.1.5 provides adjustments for braced wall line spacings of 30 and 35 feet for structures in SDC D_0, D_1 or D_2. Section R602.10.1.5 and Table R602.10.1.5 of the 2009 IRC are provided below:

> **R602.10.1.5 Braced wall line spacing for Seismic Design Categories D_0, D_1 and D_2.** *Spacing between braced wall lines in each story shall not exceed 25 feet (7620 mm) on center in both the longitudinal and transverse directions.*
>
> **Exception** *In one-and two-story buildings, spacing between two adjacent braced wall lines shall not exceed 35 feet (10 668 mm) on center in order to accommodate one single room not exceeding 900 square feet (84 m2) in each dwelling unit. Spacing between all other braced wall lines shall not exceed 25 feet (7 620 mm). A spacing of 35 feet (10 668 mm) or less shall be permitted between braced wall lines where the length of wall bracing required by (IRC) Table R602.10.1.2(2) is multiplied by the appropriate adjustment factor from (IRC) Table R602.10.1.5), the length-to-width ratio for the floor/roof diaphragm does not exceed 3:1, and the top plate lap splice face nailing shall be twelve 16d nails on each side of the splice.*

Note that IRC Section R602.10.1.5 actually contains **two** separate exceptions to the maximum braced wall line spacing of 25 feet.

- The first permits an increase to 35 feet for a single room up to 900 square feet where the spacing between all other braced wall lines does not exceed 25 feet.

- The second permits an increase up to 35 feet providing the length **of the braced wall lines so spaced** has been increased in accordance with IRC Table R602.10.1.5, AND the aspect ratio of the floor above or roof diaphragm does not exceed 3:1, AND the top plate, **of the walls running between the braced wall lines at this location,** is spliced as specified.

TABLE 3.2

Adjustments of Bracing Length for Braced Wall Lines greater than 25 feet[a,b]

IRC Table R602.10.1.5

Braced Wall Line Spacing (feet)	Multiply Bracing Length in IRC Table R602.10.1.2(2) by:
25	1.0
30	1.2
35	1.4

For SI: 1 foot = 304.8 mm
a. Linear interpolation is permitted.
b. When a braced wall line has a parallel braced wall line on both sides, the larger adjustment factor shall be used.

Remember that detached one- and two-family dwellings in SDC C and below, and townhouses in SDC A and B, are exempt from the seismic provisions in the IRC in accordance with the exception under IRC Section R301.2.2.

CHAPTER 4

Loads and Limits of the International Residential Code

Because the International Residential Code (IRC) is a prescriptive code, certain limitations are placed on its use. Geographic limitations include areas in Seismic Design Categories (SDC) A–D$_2$ and regions with wind speeds of less than 110 mph (miles per hour), 100 mph in hurricane-prone regions.

In addition to the above limitations, Chapter 3 of the IRC includes a number of geometric limitations on the size, shape, number of stories, and architectural features of the home or townhome. It is not feasible for a prescriptive code to cover every possible combination of wall layouts, cantilevers, large wall line offsets, three-sided structures, and split-level floor plans, so these limitations are necessary to prevent the IRC from being applied to homes having configurations to which conventional building practices cannot safely apply. Structures or elements of structures that exceed these limitations must be designed in accordance with the International Building Code (IBC) or other referenced standards.

Summary

The scope of the IRC
(What to do when details of the structure go beyond the IRC)

For residences outside of the geographic or geometric scope of the IRC, the designer can use the structural provisions of the IBC. Not only may the IBC be used to completely design residential structures, it also may be used to design any part or portion of the structure that is outside of the scope of the IRC. This means that the IRC and IBC can be used in combination to design a structure. This is clearly permitted by the following code sections:

> **R104.11 Alternative materials, design and methods of construction and equipment.**
> *The provisions of this code are not intended to prevent the installation of any material or to prohibit any design or method of construction not specifically prescribed by this code, provided that any such alternative has been approved. An alternative material, design or method of construction shall be approved where the building official finds that the proposed design is satisfactory and complies with the intent of the provisions of this code, and that the material, method or work offered is, for the purpose intended, at least the equivalent of that prescribed in this code. Compliance with the specific performance-based provisions of the International Codes in lieu of specific requirements of this code shall also be permitted as an alternate.*

Note: IRC Section R104.11 is also the enabling language typically used to evaluate alternative products via evaluation reports, such as those found at www.icc-es.org.

> **R301.1.3 Engineering Design.** *When a building of otherwise conventional construction contains structural elements exceeding the limits of Section R301 or otherwise not conforming to this code, these elements shall be designed in accordance with accepted engineering practice. The extent of such design need only demonstrate compliance of nonconventional elements with other applicable provisions and shall be compatible with the performance of the conventional framed system. Engineering design in accordance with the International Building Code is permitted in all buildings and structures, and parts thereof, included in the scope of this code.*

Note: from IRC Section R301.1.3, it is clear that only the element or elements that fall outside the scope of the IRC must be *designed*. It is reasonable to expect that portions of the structure that provide necessary support for these elements must also be checked to insure compatability.

In addition to using the IBC to design parts or portions of the structure, IRC Section R301.1.1 Alternative Provisions of the IRC permits the use of the following for wood construction:

- American Forest and Paper Association (AF&PA) Wood Frame Construction Manual (WFCM) as an alternative to the provisions in the IRC

- ICC-400 Standard on the Design and Construction of Log Structures

In IRC Section R301.2.1.1 Design Criteria, the code requires that when a wood-framed home is built in a region having a basic wind speed that exceeds the maximum speed permitted by the IRC, the design should be based on one or more of the following documents:

1. American Forest and Paper Association (AF&PA) Wood Frame Construction Manual for One- and Two-Family Dwellings (WFCM)

2. International Code Council (ICC) Standard for Residential Construction in High Wind Regions (ICC-600)

3. Minimum Design Loads for Buildings and Other Structures (ASCE-7)

Also added to the 2009 IRC in Section R301.2.1.1, immediately preceding the above three references, is the following statement:

> *The elements of design not addressed by those documents in items 1 through 4 shall be in accordance with this code.*

This statement was added for clarification, because the referenced documents only address the structural frame and other wind-load related elements of the building; the IRC is to be used for the remaining aspects of the project.

Construction systems – balloon and platform framing

The requirements of the IRC for light-frame construction apply to both balloon- and platform-framed buildings (IRC Section R301.1.2). Although balloon framing is not commonly used in multi-story residences, some jurisdictions require it at gable-end walls of single-story residences. These jurisdictions are mostly located in high wind regions of the U.S. and require balloon framing to prevent "hinging" between the end wall and the gable wall above as a result of positive or negative wind forces against the walls.

Note that Section R602.3 of the 2009 IRC requires studs other than jack studs, trimmer studs and cripple studs at openings in walls to be continuous from a support at the sole plate to a support at the top plate to resist loads perpendicular to the wall. The support must be a foundation or floor, ceiling or roof diaphragm.

The studs in balloon-framed walls extend from the lower story sill plate to the upper story top plate. Upper floors, when present, are typically supported by a ledger attached to or notched into the the sides of the wall studs. In platform-framed walls, the studs extend from the sill plate to the top plate for each story, and upper floors are built above lower story walls. *FIGURE 4.1* illustrates the difference between balloon and platform frame construction.

FIGURE 4.1

Balloon and platform framing

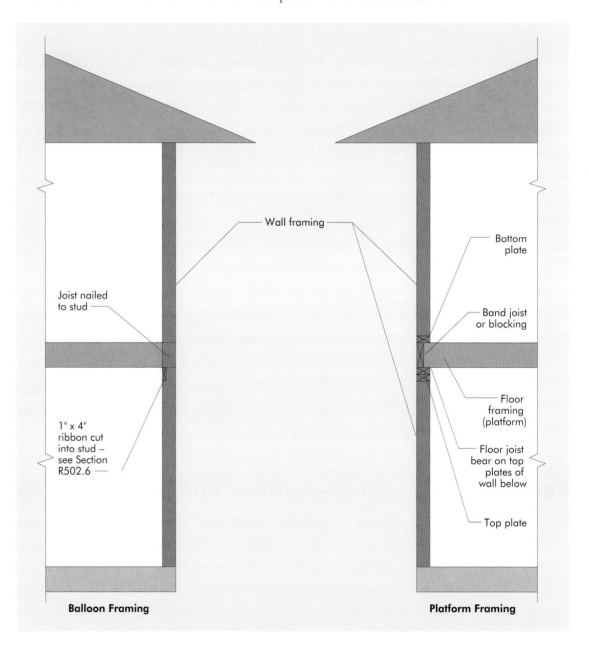

Wall framing

Joist nailed to stud

1" x 4" ribbon cut into stud – see Section R502.6

Balloon Framing

Bottom plate

Band joist or blocking

Floor framing (platform)

Floor joist bear on top plates of wall below

Top plate

Platform Framing

Story height limits

IRC Section R202 defines story height as:

> **Height, story.** *The vertical distance from top to top of two successive tiers of beams or finished floor surfaces; and, for the topmost story, from the top of the floor finish to the top of the ceiling joists or, where there is not a ceiling, to the top of the roof rafters.*

Story-height limits are provided in IRC Section R301.3 and are illustrated and discussed below. **FIGURE 4.2** shows story-height limits for light-framed wood, structural insulated panels, and cold-formed steel construction. **TABLE 4.1** provides a summary of wall heights permitted for wood-frame construction in the IRC.

FIGURE 4.2

Story height measurement for wood, structural insulated panels, and cold-formed steel framing

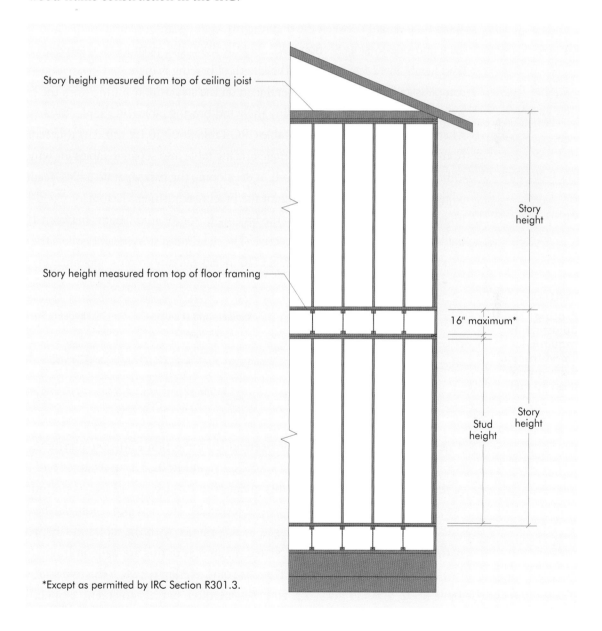

Story height measured from top of ceiling joist

Story height measured from top of floor framing

Story height

16" maximum*

Stud height

Story height

*Except as permitted by IRC Section R301.3.

Stud Height	Floor Framing Depth	Maximum Story Height
10 feet	Varies	11 feet 7 inches
12 feet[b]	Varies	13 feet 7 inches

a. Laterally unsupported bearing wall heights.
b. Wall stud clear height shall be permitted to be increased to 12 ft provided the length of bracing required by IRC Table R602.10.1.2(1) is increased by multiplying by a factor of 1.10, and the length of bracing required by IRC Table R602.10.1.2(2) is increased by multiplying by a factor of 1.20 in accordance with IRC Section R301.3, Item 1, Exception.

For wood framed-residences, the stud height for a laterally unsupported bearing wall is limited to 10 feet by IRC Table R602.3(5). Studs are laterally supported at the top and bottom by connections to diaphragms, such as a floor or roof. However, studs are not typically laterally supported along their height between the diaphragms, ceilings and/or foundations, and can deflect front-to-back and side-to-side. Section R301.3 of the 2009 IRC permits increasing laterally unsupported stud height to 12 feet, provided the length of bracing is increased by an adjustment factor to compensate for the increased lateral load on the structure resulting from the latter stud walls. The adjustment factor is 1.1 when using the wind bracing table (IRC Table R602.10.1.2(1)) and is 1.2 when using the seismic bracing table (IRC Table R602.10.1.2 (2)). The different factors resulted from separating the previously single bracing table into separate tables for wind and seismic forces (different methodologies were used in developing the two separate tables, resulting in different adjustment factors). See **TABLE 4.2** for the bracing adjustment factors for wood-framed walls. Using linear interpolation to determine the bracing factor for stud heights between 10 and 12 feet is justifiable from an engineering perspective. The maximum story height equals the stud height (12 foot maximum) plus floor framing height (16 inches maximum plus floor sheathing thickness).

Bracing Criteria	Wall Height (Feet)	Bracing increase factor
Wind Bracing	10 or less	1.0
	>10 to 12	1.1
Seismic Bracing	10 or less	1.0
	>10 to 12	1.2

A new story-height provision has been added to Section R301.3 to clarify the use of either of the two stud height tables (IRC Tables R602.3(5) and R602.3.1) in Chapter 6 of the 2009 IRC. Both of these tables address stud heights greater than 10 feet, or 12 feet as permitted with increased bracing (IRC Section R301.3, Item 1, Exception). The new provision clarifies that individual stud or wall height can be increased in accordance with the tables in Chapter 6, providing the permitted story height is not exceeded. Examples of this are balloon framed gable end walls and two-story entrance foyer walls. There is some overlap between the two stud wall tables (IRC Table R602.3(5) and R602.3.1) and they are often not in agreement. Hopefully this will be clarified in the future. At this point, story heights are limited by IRC Section R301.3 and either of the stud-height tables in Chapter 6 of the 2009 IRC may be used for entrance foyer walls or balloon-framed gable end walls.

Structures built to the provisions of the IRC may have sections of walls that exceed the prescribed wall heights of IRC Section R301.3. As long as wall sections occur between quali-fied bracing units (21-foot spacing measured between adjacent edges), IRC Tables R602.3(5) and R602.3.1 permit the use of longer framing to allow for entrance foyers and gable ends.

In addition to the overall story height, IRC Section R301.3 also limits the floor framing depth to a maximum of 16 inches. However, a new provision states that floor framing depths greater than 16 inches may be used as long as the overall story height (wall height + floor framing depth + floor sheathing thickness) does not exceed 11 feet 7 inches. This means that if 24-inch deep floor framing is used, the maximum stud/wall height would be 9 feet 7 inches (11 feet 7 inches - 24 inches = 9 feet 7 inches). The 2009 IRC permits the use of any depth of floor framing provided the combined height of the studs and floor framing does not exceed 11 feet 7 inches. This new provision in IRC Section R301.3 is reproduced below:

> *Individual walls or wall studs shall be permitted to exceed these limits as permitted by Chapter 6 provisions, provided story height does not exceed 11 feet 7 inches (3531 mm). An engineering design shall be provided...*

Note that the intention of this provision is to permit the floor framing to be greater than 16 inches in depth as long as the stud height is adjusted downward accordingly. This does not preclude the use of the provision in IRC Section R301.3, Item 1, Exception, that permits the increase of the nominal stud height from 10 feet to 12 feet providing that wall bracing is increased. Because the 11-feet 7-inch maximum story height is based on a 10-foot stud height (10-foot nominal stud height + 1-foot 4-inch floor framing + 3-inch allowance for floor sheathing), if a 12-foot stud height is used based on IRC Section R301.3, Item 1, Exception, the story height is permitted to be increased to a maximum of 13-feet 7-inches (12-foot nominal stud height + 1-foot 4-inch floor framing + 3-inch allowance for floor sheathing). Remember, the intent of this section is not to limit the stud height, it is to permit the increase in floor framing depth to over 16 inches as long as the stud height is reduced accordingly.

Question: Can a floor depth greater than 16 inches be used as long as I do not exceed my maximum story height of 136 inches (120 + 16)?

Answer: Yes. From an engineering perspective, the load contributed by a wall 136 inches high is essentially the same regardless of how much of the 136 inches in overall height is stud length and how much is floor-framing depth. The new provision added to IRC Section R301.3 permits just such an adjustment. The floor framing can be increased as long as the story height does not exceed 11 feet 7 inches (139 inches). Therefore, for this example, using a 24-inch-deep floor joist would be permissible as long as the stud length is 115 inches or less (139 inches – 24 inches = 115 inches).

Note: The story height can be increased by an additional 2 feet if the bracing length is correspondingly increased per IRC Section R301.1, Exception.

Design criteria – local design requirements

The ICC recommends that every jurisdiction that adopts the IRC fill out IRC Table R301.2(1), shown in **TABLE 4.3**. The completed table and related information can be obtained from the jurisdiction where the building will be constructed.

TABLE 4.3

Local jurisdictions must specify climatic and geographic criteria

IRC Table R301.2(1) Climatic and geographic design criteria

GROUND SNOW LOAD	WIND DESIGN		SEISMIC DESIGN CATEGORY[f]	SUBJECT TO DAMAGE FROM			WINTER DESIGN TEMP[h]	ICE BARRIER UNDER-LAYMENT REQUIRED[h]	FLOOD HAZARDS[g]	AIR FREEZING INDEX[i]	MEAN ANNUAL TEMP[i]
	Speed (mph)[d]	Topo-graphic Effects[k]		Weath-ering[a]	Frost Line Depth[b]	Termite[c]					

Footnotes not shown for clarity.

Wind loads

IRC Section R301.2.1 of the code defines the external wind pressures that components of a structure must be able to resist. The following excerpt identifies how design load performance requirements for specific exterior components are to be determined:

> *"…Where loads for wall coverings, curtain walls, roof coverings, exterior windows, skylights, garage doors and exterior doors are not otherwise specified, the loads listed in Table R301.2(2) adjusted for height and exposure using Table R301.2(3) shall be used to determine design load performance requirements for wall coverings, curtain walls, roof coverings, exterior windows, skylights, garage doors and exterior doors…"*

When wind acts on the exterior components of a building, the resulting load is transferred to the main structural elements of the building. In the IBC, the exterior components are called *components and cladding*, and the main structural elements are called the *main wind-force-resisting system*. In the IRC, main wind-force-resisting systems are treated prescriptively via specified minimum roof and floor sheathing, attachment schedules and required wall bracing. As such, by selecting the appropriate levels of bracing specified in IRC Section R602.10 and following the limitations specified in Chapter 3 of the IRC, the user will meet the necessary requirements for the main wind-force-resisting system without further design consideration.

IRC Table R301.2(2) gives the component and cladding pressure requirements that must be met by applicable exterior components. In the table, pressure values are given for basic wind speeds in 5-10 mph increments, based on wind speed, roof slope and effective wind area for the various zones of the structure (as defined in IRC Figure R301.2(7)). The zones are the building locations to which the pressure values are applied. Zones 1, 2 and 3 apply to roofs. Zones 4 and 5 apply to walls. The footnote to IRC Figure R301.2(7) specifies a 4-foot dimension for the width of the end zones at the corners and edges of the structure. Buildings having roof slopes exceeding the maximum slopes must be engineered using the IBC, per IRC Section R301.1.

The term *effective wind area*, used in IRC Table R301.2(2), is defined in Footnote a of the table.

The effective wind area shall be equal to the span length multiplied by an effective width. This width shall be permitted to be not less than one-third the span length. For cladding fasteners, the effective wind area shall not be greater than the area that is tributary to an individual fastener.

Example: What is the effective wind area for a piece of 4-foot by 8-foot wall sheathing applied vertically on an 8-foot-tall wall? The effective wind area is calculated by multiplying a height times a width. Since the wall studs vertically span 8 feet (effective height), the effective width is one-third the vertical span of the stud, which is 8 feet ÷ 3 = 2.67 feet. The effective wind area need not be less than the vertical stud span (effective height) times the effective width, which is 8 feet x 2.67 feet = 21.3 square feet. Thus, using an approximate effective wind area of 20 square feet would be reasonable for use in IRC Table R301.2(2).

IRC Table R301.2(2) shows that wall pressures for components and cladding range from +1.1 (pressure) to –26.6 pounds per square foot (suction) for 85 to 105 mph basic wind speed in Wind Exposure Category B, Zone 5. The pressure from this table is then modified by a height and exposure factor from IRC Table R301.2(3) to determine the design pressure. The exposure categories range from A to D and are defined in IRC Section R301.2.1.4 (discussed later in this chapter). The exterior wall cladding components (wall coverings, curtain walls, roof coverings, exterior windows, skylights, doors) must have sufficient capacity to withstand these wind pressures.

Component pressure capacities can be obtained from the product manufacturers or appropriate trade associations. It cannot be assumed that a product meets the code pressure requirements just because the product is commonly used. Given the poor performance of a number of siding and sheathing products during the 2003 to 2008 storm seasons in the Midwest, it appears that many commonly used siding and sheathing products have insufficient capacity to meet the wind pressure resistance requirements in some parts of the U.S.[a,b,c] Some exterior components may require additional fastening or closer support spacing to meet the requirements of this section. An example of this is IRC Table R602.3(3). This table requires larger nails and restricts stud spacing for wood structural panel sheathing for certain wind speeds and Exposure Categories within the scope of the IRC.

a. AAWA. *Reconnaissance of May 4 and 8, 2003 Kansas-Missouri Tornadoes – A Preliminary Synopsis.* The Wind Engineer – Newsletter of American Association for Wind Engineers. May 2003. American Association for Wind Engineers, Fort Collins, CO, 80522-0161. www.aawe.org

b. Zeno Martin, Bryan Readling. *Midwest Tornadoes – Performance of Wood-Framed Structures.* Wood-Design Focus – A Journal of Contemporary Wood Engineering. Winter 2008. Volume 18, Number 4. Forest Products Society, Madison, WI, 53705-2295.

c. APA – The Engineered Wood Association. *Missouri Tornadoes – Structural performance of Wood-Framed Buildings in the Tornadoes of Southwestern Missouri.* 2005. Form No. SPE-1118. Tacoma, WA. www.apawood.org

The example below shows how to determine the component and cladding wall pressures.

> **Example:** What is the "worst case" wind pressure for the wall sheathing panel in the previous wind area calculation example?
>
> The panel is in a single-story home and located in a 90 mph wind zone, Exposure Category B. The worst case for a wall panel is placement at the corner of the building. This is Zone 5 (see IRC Figure R301.2(7)). From IRC Table R301.2(2) (Wall, Zone 5, 20 square feet, 90 miles per hour), values of +13.9 psf (pounds per square foot) and −18.2 psf are shown. The height and exposure adjustment from IRC Table R301.3.2(3) is 1.0. The positive (+) value is for wind load acting from the outside toward the inside of the house (pushing the panel toward the wall studs). The negative (−) value is the wind load pulling the panel away from the studs, creating suction.
>
> **Answer:** The "worst case" wind suction pressure is 18.2 psf. The exterior components of the structure must be selected to withstand this wind pressure per IRC Section R301.2.1.

Notice that there are two wind pressure values given in the example above. This is because, as the orientation of the wind changes with respect to the structure, the wind can cause either suction (- numbers) or positive pressures (+ numbers) on the structure. For example, a wind blowing on the north side of a structure causes a positive pressure on the north side and simultaneously causes negative pressure, or suction, on the south side. As wind may blow from any direction, both pressures must be considered when sizing and selecting an exterior wall element.

Siding, sheathing and siding/sheathing combinations must be strong and stiff enough to resist positive wind pressures. They must also be sufficiently attached to prevent fastener pull out of the framing or pull through of the fastener head to avoid the result of negative wind pressures sucking cladding materials off of the framing.

Construction in high wind areas (R301.2.1.1)

If the wind speed limit for the IRC is exceeded (wind speeds of 110 mph or more, or 100 mph or more in hurricane-prone regions) design documents listed in this section must be used, as shown in *TABLE 4.4*.

Section R301.2.1.1 of the 2009 IRC includes a new reference source for high wind design, the *ICC Standard for Residential Construction in High Wind Regions, ICC-600*. This publication provides both engineered and prescriptive solutions for residential and residential-type structures in wind speed areas outside of the scope of the IRC. In many cases, it provides prescriptive solutions to problems that formerly required the services of a registered design professional. This publication is in the process of being approved as an ANSI standard.

TABLE 4.4

Applicable design standards

IRC Section R301.2.1.1 Design criteria

Basic Wind Speed	Applicable Document
Less than 110 mph	IRC
100 mph or more in Hurricane-Prone Regions[a]	See R301.2.1.1 for design reference standards
110 mph or more	

a. IRC Section R202 – Definitions: Hurricane-Prone Regions. Areas vulnerable to hurricanes, defined as the U.S. Atlantic Ocean and Gulf of Mexico coasts where the basic wind speed is greater than 90 mph (40 m/s), and Hawaii, Puerto Rico, Guam, Virgin Islands, and American Samoa.

Exposure Category (R301.2.1.4)

Exposure Category takes into account the shielding effects on wind load from other buildings, natural topography and vegetation. Adjacent buildings can deflect wind, thus reducing the pressure on the building. Tall vegetation, such as trees and terrain, with numerous, closely spaced obstructions (the size of single-family houses or larger) can also reduce the effect of wind. Topographical features, such as flat and open areas, are associated with higher Exposure Categories because there are no obstructions to reduce the effects of the wind. The code defines four Wind Exposure Categories, ranging from Exposure A (the lowest wind pressures) to Exposure D (the highest wind pressures).

Exposure Category B is the assumed Wind Exposure Category (R301.2.1.4, Item 2), unless otherwise directed by IRC Table R301.2(1) in your jurisdiction's adopted IRC.

See **FIGURES 4.3** through **4.7**, along with the Exposure Category descriptions in IRC Section R301.2.1.4, to determine the appropriate wind Exposure Category for a building's location.

FIGURE 4.3

**Exposure
Category A**

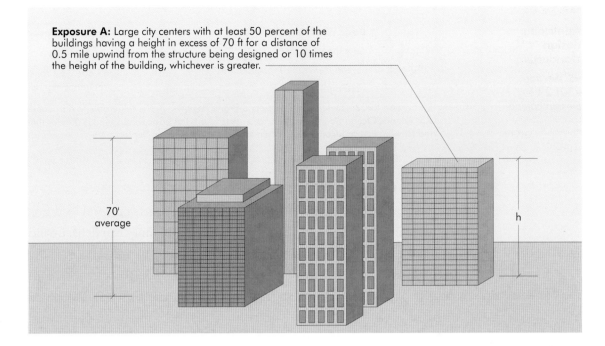

Exposure A: Large city centers with at least 50 percent of the buildings having a height in excess of 70 ft for a distance of 0.5 mile upwind from the structure being designed or 10 times the height of the building, whichever is greater.

70'
average

h

FIGURE 4.4

**Exposure
Category B**

*Shall be assumed
unless the
site meets the
definition of
another type
exposure*

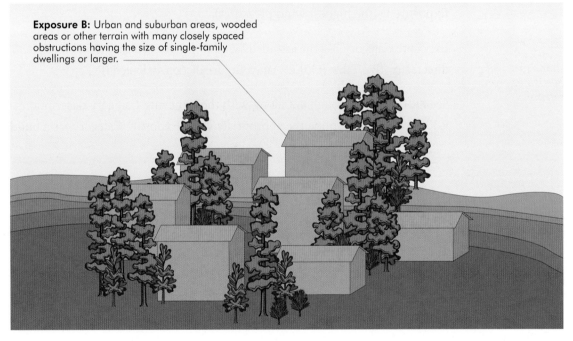

Exposure B: Urban and suburban areas, wooded areas or other terrain with many closely spaced obstructions having the size of single-family dwellings or larger.

FIGURE 4.5

Exposure Category C

This category includes open country, grasslands and shorelines in hurricane-prone regions

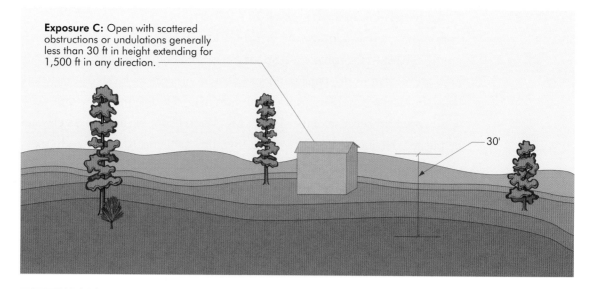

Exposure C: Open with scattered obstructions or undulations generally less than 30 ft in height extending for 1,500 ft in any direction.

30'

FIGURE 4.6

Exposure Category C (continued)

This category includes open country, grasslands and shorelines in hurricane-prone regions

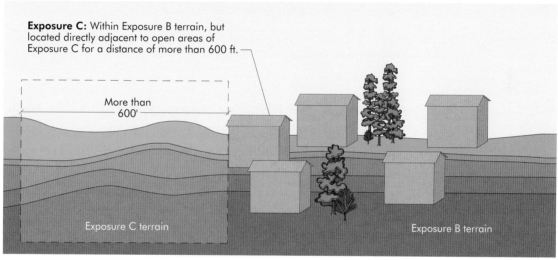

Exposure C: Within Exposure B terrain, but located directly adjacent to open areas of Exposure C for a distance of more than 600 ft.

More than 600'

Exposure C terrain

Exposure B terrain

FIGURE 4.7

Exposure Category D

This category includes inland waterways, the Great Lakes, and coastal areas of California, Oregon, Washington and Alaska

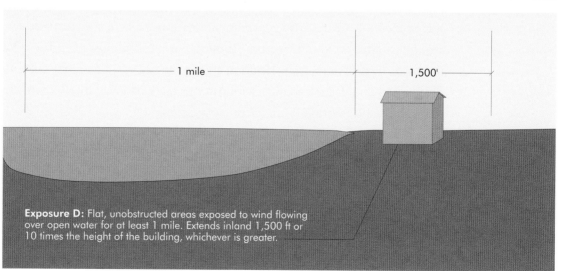

1 mile

1,500'

Exposure D: Flat, unobstructed areas exposed to wind flowing over open water for at least 1 mile. Extends inland 1,500 ft or 10 times the height of the building, whichever is greater.

Topographic wind effects (R301.2.1.5)

This section, a new addition to the 2009 IRC, recognizes that a building on level terrain will not be subject to the same wind speed as the same building on elevated terrain, even when both are located in the same wind speed zone. A building located at or near the top of a hill, ridge, or escarpment of a certain minimum height and geometry will be subject to an effective wind speed-up because of the slope. This is significant because these elevated buildings may be pushed outside of the scope of the IRC. IRC Table R301.2.1.5.1 addresses topographic wind effects and provides details on the magnitude of the wind speed-up effect. The IRC also includes a number of figures that illustrate the use of the provisions and tables. For example, a building in a 90 mph basic wind speed area that sits on the top of a hill with an average slope of 0.20 may have to be designed for a 120 mph wind, and thus will fall outside of the scope of the IRC.

Note the first sentence of IRC Section R301.2 as follows:

> *In areas designated in Table R301.2(1) as having local historical data documenting structural damage to buildings caused by wind speed-up at isolated hills, ridges, and escarpments that are abrupt change from the general typography of the area, topographic wind effects shall be considered...*

The information in IRC Table R301.2(1) is provided by the local building official, as it is their experience as to the historical impact of wind speed-up on building damage that will dictate whether topographical effects will be considered in the local jurisdiction. The house designer is encouraged to contact the local building official before considering topographical wind effects.

Seismic loads – Seismic Design Categories A, B and C

Seismic provisions (IRC Section R301.2.2)

IRC Section R301.2.2 provides the limitations applicable to structures designed for the seismic requirements of the IRC. One of the most important and most often missed provisions is that the application of the seismic requirements of the IRC is different for one- and two-family dwellings than it is for buildings with three or more attached dwelling units (townhouses). The exception to IRC Section R301.2.2 provides as follows:

> **Exception:** *Detached one- and two-family dwellings located in Seismic Design Category C are exempt from the seismic requirements of this code.*

While the exception appears to limit the exemption to Seismic Design Category (SDC) C, IRC Table R602.10.1.2(2) also exempts SDC A and B.

IRC Section R202 provides the definitions of dwelling unit, dwelling and townhouse.

Dwelling Unit: *A single unit providing complete independent living facilities for one or more persons, including permanent provisions for living, sleeping, eating, cooking and sanitation.*

Dwelling: *Any building that contains one or two dwelling units used, intended, or designed to be built, used, rented, leased, let or hired out to be occupied, or that are occupied for living purposes.*

Townhouse: *A single-family dwelling unit constructed in a group of three or more attached units in which each unit extends from the foundation to roof and with a yard or public way on at least two sides.*

In **FIGURE 4.8**, plans A and B meet the requirements of a townhouse because at least two sides of each unit have open space, or exterior walls. The plan shown in **FIGURE 4.9** does not meet the townhouse definition because the middle units do not have two sides that have "a yard or public way" (exterior walls); rather, those units only have one side that has open space. All units of a townhouse must extend from the foundation to the roof; in other words, the dwelling units cannot be stacked one above another.

FIGURE 4.8

Both plans A and B meet townhouse requirements because each unit has a yard or public way on two sides

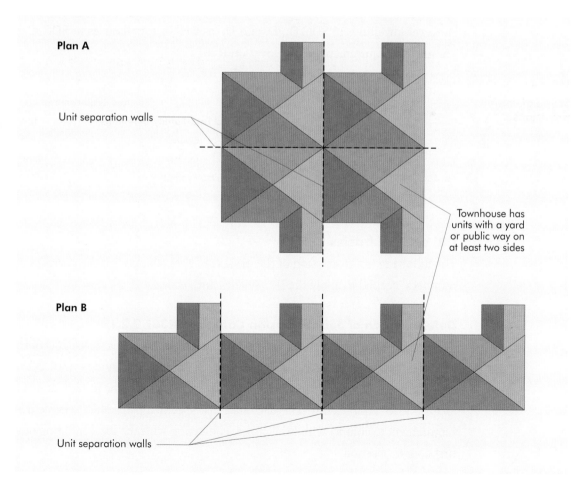

Plan A

Unit separation walls

Townhouse has units with a yard or public way on at least two sides

Plan B

Unit separation walls

FIGURE 4.9

Plan view shows a configuration not covered by the IRC

Structure does not meet requirements of a dwelling (more than 2 units) or a townhouse (middle units without a yard or public way on two sides)

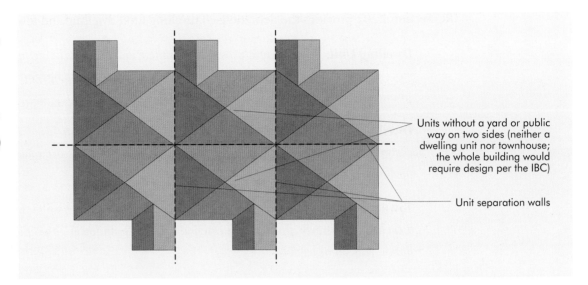

Units without a yard or public way on two sides (neither a dwelling unit nor townhouse; the whole building would require design per the IBC)

Unit separation walls

TABLE 4.5, developed from IRC Section R301.2.2, illustrates the importance of distinguishing between dwellings and townhouses. Note that dwellings located in SDC A, B and C are exempt from the seismic requirements of the IRC. This means that these dwellings are not required to comply with the seismic provisions of 301.2.2 or its subsections, or those in Chapters 4, 5, 6 and 7 of the IRC. This exception significantly reduces the construction requirements for any dwelling built in SDC A or B, and for one- and two-family dwellings built in SDC A, B and C. However, dwellings in SDC D_0, D_1 and D_2, as well as townhouses in SDC C, must comply with the seismic requirements of IRC Section R301.2.2.

TABLE 4.5

Scope of seismic provisions

Seismic Design Categories	Seismic Requirements	
	One- and Two-Family Dwellings	Townhouses
A – B	Exempt	Exempt
C	Exempt	No Exemption
$D_0 – D_2$	No Exemption	No Exemption

IRC seismic basics

The following is a discussion of the basic concepts, along with the corresponding IRC section numbers, needed to apply the IRC seismic provisions.

Determination of Seismic Design Category (R301.2.2.1)

The Seismic Design Category (SDC) can be found in IRC Table R301.2(1) of the locally adopted IRC (see **TABLE 4.3**). If not available, the maps shown in IRC Figures R301.2(2) may be used, or contact the local building official for assistance. Note that some jurisdictions may amend the local code to require a different SDC than that specified by the IRC seismic maps. Because of the requirement to comply with local codes, conferring with the local building official is always a wise first course of action.

Alternative determination of Seismic Design Category (R301.2.2.1.1)

The SDCs used in the IRC, as well as in the IBC, assume soil site class D when the soil properties are not known in sufficient detail to determine the site classification. If soil conditions are known to be other than site class D, the SDC for the site can be determined using the 2009 IBC Section 1613.5.2 or ASCE 7-05 Section 11.4.2 to determine S_{DS} (spectral short-period response acceleration parameter) and IRC Table R301.2.2.1.1 to find the corresponding SDC. The local building official is also a good source of information on local soil types.

Alternative determination of Seismic Design Category E (R301.2.2.1.2)

In some cases, a building located in SDC E, according to IRC Figure R301.2(2), may be reclassified as SDC D_2. This would put the building within the scope of the IRC, thus eliminating the need for a full design. These cases are described in IRC Section R301.2.2.1.2. This alternate determination is possible because of necessary compromises made in the development of the seismic maps used in the IRC. The work done to simplify the maps made the SDC E contours more conservative than their IBC counterparts. The alternate determination provisions of IRC Section R301.2.2.1.2 are permitted to compensate for this conservatism.

Seismic Design Category C (R301.2.2.2)

The limitations provided in this section apply to residential structures in SDC C and above. This section details the scoping limitations aimed at restricting architectural features that may make a structure unsuitable for the prescriptive methodology used in the IRC. As noted previously, detached one- and two-family dwellings in SDC C are exempt from the seismic requirements of the code, so these provisions only apply to townhouses.

Weight of materials (R301.2.2.2.1)

The seismic loads on a building are related to building weight (mass). The greater the building weight, the greater the seismic load (Force = Mass x Acceleration). During ground motion/acceleration, the structural parts of a building must resist the lateral forces to prevent collapse. Therefore, the IRC sets limits on the weights of floors, walls, ceiling and roofs. *TABLE 4.6* lists the maximum building weights defined by this section for wood frame construction. This section does permit the increase of some building weights, provided the amount of wall bracing is increased. Increasing the amount of wall bracing makes the building stronger, stiffer and more capable of resisting the greater seismic loads that result from the greater building mass (see *FIGURE 4.10*).

FIGURE 4.10

Maximum dead load weights

IRC Section R301.2.2.2.1

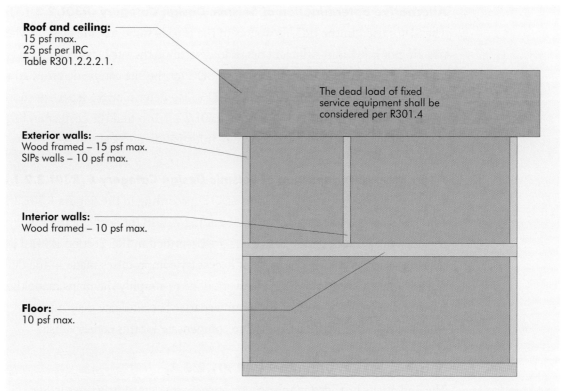

Roof and ceiling:
15 psf max.
25 psf per IRC Table R301.2.2.2.1.

The dead load of fixed service equipment shall be considered per R301.4

Exterior walls:
Wood framed – 15 psf max.
SIPs walls – 10 psf max.

Interior walls:
Wood framed – 10 psf max.

Floor:
10 psf max.

Note: The allowable load given for roofs is the weight of the roof based on the horizontal projection of the roof. This is the total weight of the roof divided by the area of the roof in plan view.

TABLE 4.6

Bracing adjustment factors based on weights of construction materials

Per IRC Table R301.2.2.2.1

		Dead Loads and Wall Bracing Adjustment Factors		
Component	**Story Location**	**Maximum Dead Load Permitted for Standard Wall Bracing (psf)**	**Maximum Dead Load Permitted with Wall Bracing Adjustment (psf)**	**Bracing Panel Amount Adjustment Factor[a]**
Floor		10	N/A	N/A
Roof/Ceiling Dead Load For Wall Supporting the Roof Only		15[b]	25[b]	1.2
Roof/Ceiling Dead Load For Wall Supporting the Roof Plus One or Two Stories		15[b]	25[b]	1.1

a. This factor is to be applied to the required bracing amount (length) of the wall line. See IRC Table R602.10.1.2(2).
b. The allowable load given for roofs is the weight of the roof based on the horizontal projection of the roof. This is the total weight of the roof divided by the area of the roof in plan view.

Example: Consider a braced wall that has only a roof above it. The roof is heavy tile with a total dead load of 20 psf measured on the horizontal projection. Since 20 psf exceeds the 15 psf limit, but is less than 25 psf, the amount of bracing required per IRC Table R602.10.1.2(2) must be increased by a factor of 1.2. While not specifically provided for in the code provisions, interpolation between 1.0 and 1.2 is justifiable from an engineering perspective.

Example: A home designer draws up a set of plans for a two-story townhouse. After working through the bracing section of the IRC, the designer determines that the structure, based on the seismic bracing requirements, requires each exterior wall on the first story to have 18 feet of bracing and the second story to have 10 feet of bracing. The designer assumes that, like most townhouses in the area, the structure will have a typical light-weight wood roof (roof/ceiling dead load of 15 psf or less); however, the owner later asks the designer to consider a slate roof. This roofing material adds 6 psf to the weight to the horizontal projection of the roof. How does this impact the amount of bracing needed for the first and second stories?

With a slate roof, the weight of the roof/ceiling becomes 21 (15+6) psf. From *TABLE 4.6*, 21 psf is less than the maximum dead load permitted (25 psf). This means that a bracing length adjustment factor of 1.1 is required for the first story and 1.2 for the second story.

First story   18 feet x 1.1 = 19.8 feet of bracing required

Second story 10 feet x 1.2 = 12 feet of bracing required

Note that the provisions of IRC Table R301.2.2.2.1 (*TABLE 4.6*) appear on the surface to be counterintuitive. According to the table, the impact of increased roof seismic weight appears greater for a single story than for a two-story structure. From an engineering perspective, this is correct. When there is just a roof over a single-story structure, any increase in the seismic weight of the roof will have approximately the same increase in the required bracing on the walls of the first story. For example, a 50 percent increase in roof seismic weight causes the loads on the braced wall below to be increased by approximately the same amount.

However, when there is a full story above the first story walls, any increase in the seismic weight on the roof of a given percentage when combined with the seismic weight of the second floor results in a smaller percentage increase to the walls on the first floor.

What the table indicates is that with a roof-ceiling dead load increase from 15 to 25 percent, the walls supporting the roof only experience a 20 percent increase in load on the braced wall lines. For the walls supporting the roof <u>and</u> second floor, this is only a 10 percent increase.

Stone and masonry veneer

Another change in the 2009 IRC is the relocation of the wall bracing provisions for stone or masonry wall veneer from Chapter 7 of the IRC to Chapter 6, where the other bracing provisions reside. These bracing provisions are now included in IRC Section R602.12. See **CHAPTER 10** for a summary of the stone and masonry veneer bracing requirements.

Building irregularities (IRC Section R301.2.2.2.5)

This section of the code defines building shapes that are irregular. It is important to note that this section applies only to structures located in SDC C-D$_2$, excluding one- and two-family dwellings in SDC C. The IRC seismic provisions assume that a building has a relatively uniform shape. If the building deviates from the assumed uniform shape, the building or portion of the building is considered irregular. IRC Section R301.2.2.2.5 defines these irregularities and places limits on them for purposes of inclusion within the IRC.

The problem with irregular buildings is that the irregular shape of the structure can cause the lateral loads to shift around in the building in unusual and difficult-to-predict load paths. Such structures are difficult to design and almost impossible to effectively specify using the prescriptive provisions of the IRC. In order to permit some degree of building irregularity that the prescriptive provisions can safely accommodate, this section was developed. Those structures, or portions thereof, that are outside of these limitations are irregular shaped buildings and must be designed using the IBC to ensure that the building has adequate capacity and a load path to resist seismic forces. Buildings having a regular shape can be constructed by following the prescriptive provisions of the IRC.

Many contemporary home designs may be considered irregular by one of the seven conditions of this section. These provisions, however, include some exceptions that may permit an otherwise irregular home to be constructed by the IRC.

Irregularity #1: Braced wall panels are not in a vertical plane

The first irregularity provision requires that braced wall panels be vertically aligned in one plane.

Following this irregularity provision, an exception is listed for buildings constructed using wood light-frame construction, as shown in **FIGURE 4.11**. This exception is not applicable to other types of construction, such as masonry or concrete. According to the exception, braced wall panels do not have to be in one plane vertically if:

Setbacks or cantilevers do not exceed four times the nominal depth of the wood floor joists and ALL of the following conditions are met:

1. Floor joists are nominal 2x10 or larger and spaced not more than 16 inches on center. (Note that some engineered I-joist products are manufactured at 9-1/2-inch depths. These can be considered within the scope of this condition.)

2. The ratio of the backspan to cantilever is at least 2:1.

3. Floor joists at the ends of braced wall panels are doubled.

4. A continuous rim joist is connected to ends of all cantilevered joists.

5. Gravity loads at the end of the cantilever are limited to uniform wall and roof loads and the headers in the wall must have a span of 8 feet or less.

FIGURE 4.11

Irregularity #1: Braced wall panels with vertical irregularities

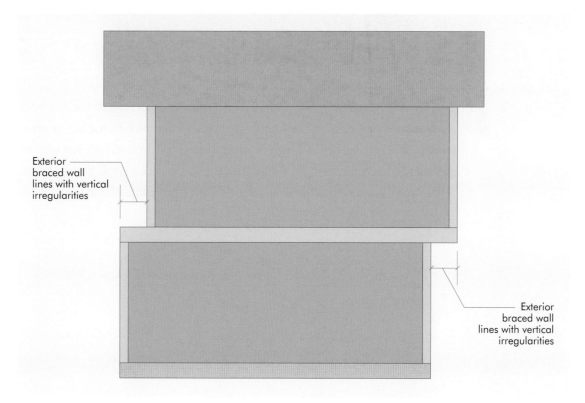

Exterior braced wall lines with vertical irregularities

Exterior braced wall lines with vertical irregularities

Irregularity #2: Braced wall line not under floor or roof

According to the second provision, a building is considered irregular when a braced wall line is not beneath (or supporting) a portion of a roof or floor. If this section of the floor or roof is not in turn supporting any braced wall panels located above, it is permitted to extend out not more than 6 feet. The exception to the irregularity is shown in ***FIGURE 4.12***.

FIGURE 4.12

Exception to Irregularity #2: Portions of floor or roof not supported are permitted to extend up to 6 feet

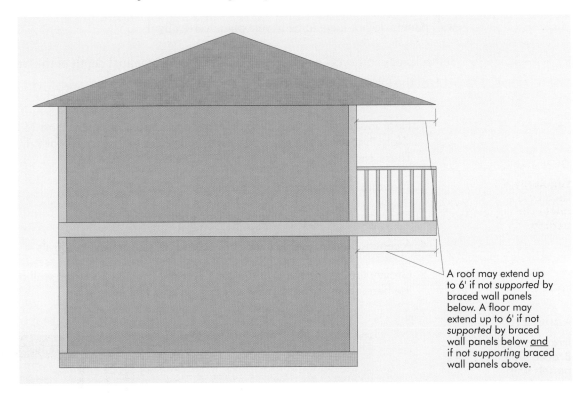

A roof may extend up to 6' if not *supported* by braced wall panels below. A floor may extend up to 6' if not *supported* by braced wall panels below <u>and</u> if not *supporting* braced wall panels above.

This second irregularity provision is different than the first, in that the first irregularity provision relates to braced wall lines being supported by floor framing. In the exception to the second irregularity provision, a roof or floor can be unsupported by a braced wall line for a maximum of 6 feet. Stone and masonry construction is outside the scope of this book. Stone and masonry veneer, which is within the scope of this book, is different from stone and masonry construction.

Irregularity #3: Braced wall panel occurs over opening below

The third irregularity provision in this section places limitations on the location of the end of braced wall panels relative to openings in the braced wall line below. In general, a braced wall panel is only permitted to extend 1 foot over an opening in a lower wall line, as shown in **FIGURE 4.13**. For wood light-frame construction, a braced wall panel is permitted to extend more than 1 foot over the opening below, provided the entire braced wall panel is not located over the opening, and the exceptions are met.

For wood light-frame wall construction, a number of exceptions apply to irregularity #3. Provided that the building width, loading conditions and framing member species meet the requirements of IRC Table R502.5(1), and the entire length of the braced wall panel does not occur over the opening, the header sizes shown in **TABLE 4.7** may be used to exempt irregularity #3.

FIGURE 4.13

Irregularity #3: Braced wall panel over openings

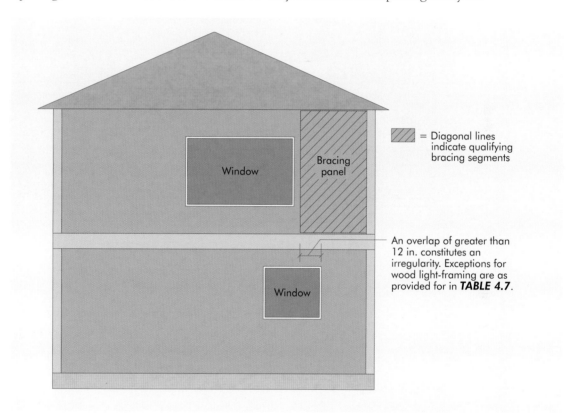

= Diagonal lines indicate qualifying bracing segments

Bracing panel

Window

Window

An overlap of greater than 12 in. constitutes an irregularity. Exceptions for wood light-framing are as provided for in **TABLE 4.7**.

TABLE 4.7

Header requirements to exempt irregularity #3

Maximum Window Opening Width Below Bracing	Minimum Header Requirements
4 ft	1 – 2x12
	2 – 2x10
6 ft	2 – 2x12
	3 – 2x10
8 ft	3 – 2x12
	4 – 2x10

Irregularity #4: Floor or roof opening limit

The fourth irregularity limits the size of the floor opening between stories or an opening in the roof. This limitation applies to all types of construction. The purpose of this restriction is to limit the size of a hole in the floor or roof diaphragm. An example of an opening between floors would be a stairway or room with a height that extends through the story above. A skylight would be considered a roof opening. The size of floor and roof openings is limited to the lesser of 12 feet or 50 percent of the least floor or roof dimension as shown in **FIGURE 4.14**.

FIGURE 4.14

Irregularity #4: Excessive hole in roof or floor sheathing

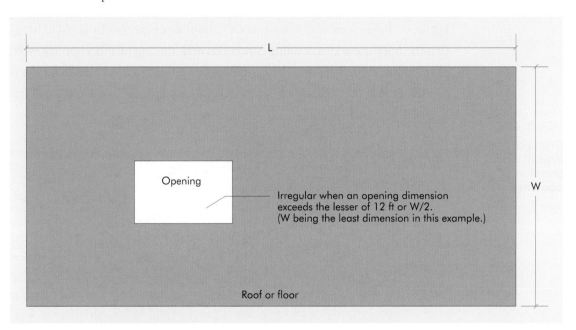

Irregularity #5: Vertical offsets in floor level

For the fifth irregularity provision, a building is considered irregular when a floor has a vertical offset as shown in *FIGURE 4.15*. For wood light-framed construction, the building is considered regular if the floor framing is lapped or tied together per IRC Section R502.6.1. For all construction types (wood, steel, etc.), vertical offsets in a floor level shall not be considered irregular if the perimeter framing of the floor is supported directly on a continuous foundation at the building perimeter.

IRC Section R502.6.1 states that joist framing must be lapped a minimum of 3 inches and face-nailed together with three 10d nails. As an alternate, a wood or metal splice with equal strength is permitted. These measures address the problems associated with floor-offset irregularities.

FIGURE 4.15

Irregularity #5: Offset in floor framing

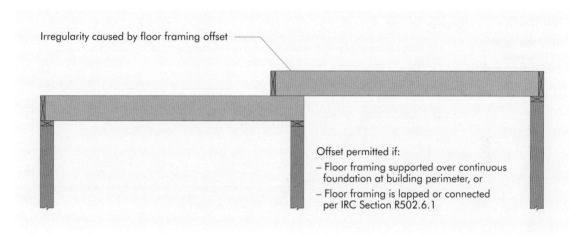

Irregularity caused by floor framing offset

Offset permitted if:
– Floor framing supported over continuous foundation at building perimeter, or
– Floor framing is lapped or connected per IRC Section R502.6.1

Irregularity #6: Braced wall lines must occur in perpendicular directions

Irregularity provision #6 simply states that if braced wall lines are not perpendicular to each other, the building must be considered irregular. **FIGURE 4.16** illustrates such an irregularity. There is no exception to mitigate this type of irregularity.

Note that an angled corner as described in IRC Section R602.10.1.3 (illustrated in **FIGURE 7.11**) is not such an irregularity, but rather an architectural feature found on a basically rectangular structure. This section permits the bracing that may be found on the angled portion to be counted towards the minimum length required for one wall. The angled corner is limited to 8 feet in length, thus minimizing its impact on the structural rectangularity of the building.

FIGURE 4.16

Irregularity #6: Braced wall lines not at right angles to each other

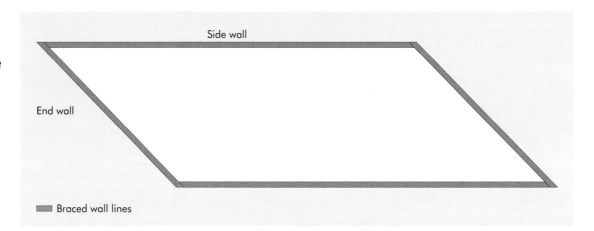

Irregularity #7: Above-grade masonry or concrete construction

The seventh irregularity provision states that wood- and steel-framed buildings braced in accordance with IRC Chapter 6 are irregular when they include above-grade masonry or concrete construction. Fireplaces, chimneys and masonry veneer are permitted, provided they are constructed in accordance with other provisions of the IRC. Note that when this irregularity exists, the entire story shall be designed, not just the portion causing the irregularity.

Additional seismic limits – Seismic Design Categories D_0, D_1 and D_2

In addition to the limitations discussed, there are additional limitations placed on residential structures in SDC D_0, D_1 and D_2: in accordance with IRC Section R301.2.2.3, buildings constructed in these SDCs are required to meet all of the provisions of buildings constructed in SDC C, plus the added requirements of this section and its seven subsections (one of which relates to bracing of wood-framed construction).

Height limitations (IRC Section R301.2.2.3.1)

This section defines the maximum number of stories permitted for light wood- and steel-framed buildings in SDC D_0, D_1 and D_2. Light wood-framed buildings are permitted to be a maximum of three stories, but only two stories in D_2 (this limit is met by reference to the seismic bracing table, IRC Table R602.10.1.2(2)). See *TABLE 4.8*.

TABLE 4.8

Height limitations (SDC D_0, D_1 and D_2)

Framing Material	Maximum Number of Stories
Wood	3[a]

a. Table R602.10.1.2(2) limits wood-framed structures in SDC D_2 to 2 stories and a cripple wall.

Stone and masonry veneer (IRC Section R301.2.2.3.2)

Anchorage requirements for SDC D_0, D_1 and D_2 are included in IRC Section R703.7. IRC Tables R602.12(1) and R602.12(2) provide bracing and story limitations for stone and masonry veneer. IRC Tables R703.7(1) and R703.7(2) provide height and story limitations.

- Bracing Limitations: IRC Tables R602.12(1) and (2)

- Height Limitations: IRC Tables R703.7(1) and (2)

- Story Limitations: IRC Tables R602.12(1) and (2) and R703.7(1) and (2)

Seismic Design Category E (IRC Section R301.2.2.4)

This new addition to the 2009 IRC, discussed previously, clarifies that structures in SDC E must be designed in accordance with the IBC, except when the structure is reclassified in accordance with IRC Section R301.2.2.1.2 as shown below:

- **R301.2.2.1.2 Alternative determination of Seismic Design Category E.** *Buildings located in Seismic Design Category E in accordance with Figure R301.2(2) are permitted to be reclassified as being in Seismic Design Category D_2 provided one of the following is done:*

  1. *A more detailed evaluation of the seismic design category is made in accordance with the provisions and maps of the International Building Code. Buildings located in Seismic Design Category E per Table R301.2.2.1.1, but located in Seismic Design Category D per the International Building Code, may be designed using the Seismic Design Category D_2 requirements of this code.*

  2. *Buildings located in Seismic Design Category E that conform to the following additional restrictions are permitted to be constructed in accordance with the provisions for Seismic Design Category D_2 of this code:*

    2.1. *All exterior shear wall lines or braced wall panels are in one plane vertically from the foundation to the uppermost story.*

    2.2. *Floors shall not cantilever past the exterior walls.*

    2.3. *The building meets all of the requirements of Section R301.2.2.2.5 for being considered as regular.*

Snow loads

For all construction types, 70 psf (pounds per square foot) is the maximum ground snow load permitted for buildings constructed per the IRC (R301.2.3). Buildings with a ground snow load greater than 70 psf must be designed in accordance with accepted engineering practice.

CHAPTER 5

Description of Bracing Methods

What's new for the 2009 IRC

When it comes to selecting a bracing method for a braced wall line, there are many construction and material options to choose from. The 2009 International Residential Code (IRC) offers multiple new bracing options, including bracing methods that are narrower than the traditional 4-foot length.

In the 2009 IRC, bracing methods are no longer referred to by number, such as *Method 5* for gypsum board bracing or *Method 6* for particleboard sheathing. Instead, these methods are now identified by abbreviations, such as *Method GB* for gypsum board bracing and *Method PBS* for particleboard sheathing. The ICC Ad Hoc Committee on Wall Bracing believes that this new identification method is more understandable and will make it easier to add or delete methods.

The bracing methods are now categorized into two application types: intermittent and continuous. Intermittent bracing refers to bracing methods that can be used in discrete locations along a braced wall line. Continuous sheathing bracing methods require that the whole wall line be sheathed, including above and below openings and at gable ends, if present. The continuous methods permit the use of narrower bracing panels, allowing home designers greater flexibility in selecting the size and location of doors and windows. Continuous methods addressed in the 2009 IRC also provide a substrate suitable for the attachment of nonstructural siding materials in areas exposed to higher wind speeds.

Summary

Another change in the 2009 IRC is that, for both intermittent and continuous braced wall panels, gypsum wall board must be installed on the side of the wall opposite the bracing material. There are some exceptions to this rule for the narrow bracing methods and portal frames. These new requirements and exceptions are detailed later in this chapter.

Overview

Intermittent braced wall methods

The eleven intermittent bracing methods listed in Section R602.10.2 of the 2009 IRC include eight traditional methods and two alternate methods (all permitted in the 2006 IRC) plus one new alternate method. The eight traditional methods – including let-in bracing, diagonal wood board bracing and six panel-type bracing methods – and their 2009 IRC designations are:

1. Let-in bracing (Method LIB) (formerly Method 1)

2. 5/8-inch diagonal wood boards (Method DWB) (formerly Method 2)

3. Wood structural panel (plywood or OSB) (Method WSP) (formerly Method 3)

4. 1/2-inch structural fiberboard (Method SFB) (formerly Method 4)

5. 1/2-inch interior gypsum board or gypsum sheathing (Method GB) (formerly Method 5)

6. Particleboard sheathing (Method PBS) (formerly Method 6)

7. Portland cement plaster (Method PCP) (formerly Method 7)

8. Hardboard panel siding (Method HPS) (formerly Method 8)

As the name implies, intermittent bracing methods are meant to be used intermittently along the wall line (as shown in **FIGURE 5.1**). For these methods, the minimum length required for a single braced wall panel segment ranges from 48 to 96 inches. In the 2009 IRC, slightly shorter panel lengths are permitted, but with a penalty. For example, a 42-inch particleboard sheathing (Method PBS) panel may be used as a bracing panel, but when computing the required length of bracing, it is equivalent to only 36 inches of wall bracing.

FIGURE 5.1

Example of "intermittent" Method WSP (wood structural panel) bracing panel

Intermittent bracing needs only to occur in isolated, specified locations

Three alternative methods are also classified as intermittent bracing methods because, while structurally different from the eight traditional methods, these methods are designed to be used intermittently and can be substituted for any of the traditional bracing methods on a one-for-one basis (they may be intermixed with other intermittent bracing methods along the same wall line). In the 2009 IRC, these are designated as:

1. Alternate braced wall (Method ABW) (formerly alternate braced wall panel)

2. Intermittent portal frame (Method PFH) (formerly alternate braced wall panel adjacent to a door or window opening)

3. Intermittent portal frame at garage door openings in Seismic Design Categories A, B and C (Method PFG) (new to the 2009 IRC)

Method ABW (alternate braced wall) and Method PFH (intermittent portal frame) were both included in the 2006 IRC as alternative methods. The minimum lengths for these methods range from 16 to 40 inches. Either may be substituted for any bracing method permitted in Sections R602.10.2 of the IRC. For the purposes of computing the required length of bracing, a single unit of either method is equivalent to 4 feet. Both utilize prefabricated metal hold downs, in addition to anchor bolts, to connect the bracing panel to the foundation below.

WHAT IS A HOLD DOWN?

A hold down is a prefabricated metal anchoring device that attaches the framing of a wall system to the structure below. Ultimately, the hold down load path must extend down into the foundation. The hold down prevents uplift of the studs and, thus, over-turning of the wall segment. One of a number of types is shown.

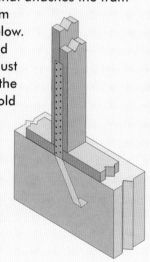

The third alternative method, Method PFG (intermittent portal frame at garage door openings in Seismic Design Categories A, B and C), is new to the 2009 IRC. This bracing method is a type of portal frame that does not require hold downs, has a minimum length of 24 inches, may only be used in SDC A, B and C, and may only be used adjacent to a garage door. For the purposes of computing the required length of bracing, a single unit of this method is equivalent to 1.5 times its actual length.

Required finish for intermittent braced wall methods

New to the 2009 IRC is a requirement that, for intermittent braced wall panels, regular gypsum wall board must be installed as an interior finish material on the side of the wall opposite the bracing material (IRC Section R602.10.2.1). Note that this requirement is for the use of standard gypsum wall board, not Method GB (gypsum board) bracing. Interior finish gypsum wall board can be attached in accordance with IRC Section R702.3.5 (7-inch on center attachment required for Method GB does not apply). Narrow wall intermittent bracing methods (AWB, PFG and PFH) are exempt from this requirement because they were developed without gypsum wall board interior finish and they are typically used adjacent to garage door openings, where gypsum wall board is not commonly used. Additionally, there are exceptions that eliminate this requirement for gypsum wall board as an interior finish material if the wall is sufficiently over-braced, per IRC Section R602.10.2.1:

> **R602.10.2.1 Intermittent braced wall panel interior finish material.** *Intermittent braced wall panels shall have gypsum wall board installed on the side of the wall opposite the bracing material. Gypsum wall board shall be not less than 1/2 inch (12.7 mm) in thickness and be fastened in accordance with Table R702.3.5 for interior gypsum wall board.*
>
> **Exceptions:**
> 1. *Wall panels that are braced in accordance with Methods GB, ABW, PFG and PFH.*
>
> 2. *When an approved interior finish material with an in-plane shear resistance equivalent to gypsum board is installed.*
>
> 3. *For Methods DWB, WSP, SFB, PBS, PCP and HPS, gypsum wall board is permitted to be omitted provided the length of bracing in Tables R602.10.1.2(1) and R602.10.1.2(2) is multiplied by a factor of 1.5.*

Adhesive attachment of intermittent braced wall methods

The provision provided below, carried over from the 2006 IRC, prohibits the use of adhesives in moderate-to-high seismic areas (SDC C, D_0, D_1 and D_2). The intent of this provision is that only sheathing *used as a part of a braced wall segment*, as described in IRC Table R602.10.2, shall not be attached with adhesives. Because it is a subsection of IRC Section R602.10.2 (intermittent braced wall construction methods), this provision is only relevant to the intermittent bracing methods as the code is currently written.

R602.10.2.2 Adhesive attachment of sheathing in Seismic Design Categories C, D_0, D_1 and D_2. *Adhesive attachment of wall sheathing shall not be permitted in Seismic Design Categories C, D_0, D_1 and D_2.*

It can also be correctly interpreted that this provision does not apply to IRC Section R602.10.2.1 (intermittent braced wall panel interior finish material). While the interior finish is a required part of the bracing panel segment, its contribution to the overall bracing strength is relatively minor, and is actually predicated on the fastener spacing used in conjunction with the adhesive attachment of gypsum wall board.

Continuous wood structural panel sheathing bracing methods

Three additional bracing methods are included in the 2009 IRC (Section R602.10.4). These methods have their roots in various provisions of the 2006 IRC, but are clarified in the new edition. Examples are provided in **FIGURE 5.13**.

1. Method CS-WSP (continuous wood structural panel sheathing) (formerly Section R602.10.5 of the 2006 IRC)

2. Method CS-G (wood structural panel adjacent to garage door openings and supporting roof loads only) (formerly IRC Table R602.10.5, Footnote b)

3. Method CS-PF (continuous portal frame) (formerly IRC Table R602.10.5, Footnote c)

Continuous structural fiberboard sheathing bracing wall method

New to the 2009 IRC is bracing Method CS-SFB (continuous structural fiberboard sheathing). It is similar to Method CS-WSP (continuous wood structural panel sheathing) except that there are restrictions on where it may be used. This method is described in a different section of the code (IRC Section R602.10.5) than Method CS-WSP and, therefore, is considered separately in this publication. However, for the purposes of comparing this bracing method with all others, it is summarized in a table format later in this chapter.

Description of intermittent bracing methods

TABLE 5.1 displays the eleven intermittent bracing methods contained in the 2009 IRC. Much of the included content, including minimum thickness information and connection requirements, is new for the 2009 IRC. Following **TABLE 5.1** are additional tables and figures that further detail the intermittent bracing methods. Although the methods are, in most cases, not detailed in paragraph form, appropriate sections of the code are referenced as needed.

TABLE 5.1

Intermittent bracing methods

IRC Table R602.10.2

Method	Material	Minimum Thickness	Figure	Connection Criteria
LIB	Let-in-bracing	1x4 wood or approved metal straps at 45° to 60° angles for maximum 16" stud spacing		Wood: 2-8d nails per stud including top and bottom plate Metal: per manufacturer
DWB	Diagonal wood boards	3/4" (1" nominal) for maximum 24" stud spacing		2-8d (2-1/2" x 0.113") nails or 2 staples, 1-3/4" per stud
WSP	Wood structural panel (See Section R604)	3/8"		Exterior sheathing: see IRC Table R602.3(3) Interior sheathing: see IRC Table R602.3(1)
SFB	Structural fiberboard sheathing	1/2" or 25/32" for maximum 16" stud spacing		1-1/2" galvanized roofing nails or 8d common (2-1/2" x 0.131") nails at 3" spacing (panel edges) at 6" spacing (intermediate supports)
GB	Gypsum board	1/2"		Nails and screws at 7" spacing at panel edges including top and bottom plates for all braced wall panel locations. For exterior sheathing nail or screw size, see IRC Table R602.3(1). For interior gypsum board nail or screw size, see IRC Table R702.3.5.
PBS	Particleboard sheathing (See Section R605)	3/8" or 1/2" for maximum 16" stud spacing		1-1/2" galvanized roofing nails or 8d common (2-1/2" x 0.131") nails at 3" spacing (panel edges) at 6" spacing (intermediate supports)
PCP	Portland cement plaster	See Section R703.6 for maximum 16" stud spacing		1-1/2", 11 gage, 7/16" head nails or 7/8", 16 gage staples at 6" spacing
HPS	Hardboard panel siding	7/16" for maximum 16" stud spacing		0.092" dia., 0.225" head nails with length to accommodate 1-1/2" penetration into studs at 4" spacing (panel edges), at 8" spacing (intermediate supports)
ABW	Alternate braced wall	See Section R602.10.3.2		See Section R602.10.3.2
PFH	Intermittent portal frame	See Section R602.10.3.3		See Section R602.10.3.3
PFG	Intermittent portal frame at garage	See Section R602.10.3.4		See Section R602.10.3.4

Method LIB (let-in bracing)

Note the maximum stud spacing is 16 inches.

Method	Material	Minimum Thickness	Figure	Connection Criteria
LIB	Let-in bracing	1x4 wood or approved metal straps at 45° to 60° angles for maximum 16" stud spacing		Wood: 2-8d nails per stud including top and bottom plate Metal: per manufacturer

FIGURE 5.2

Method LIB (let-in bracing)

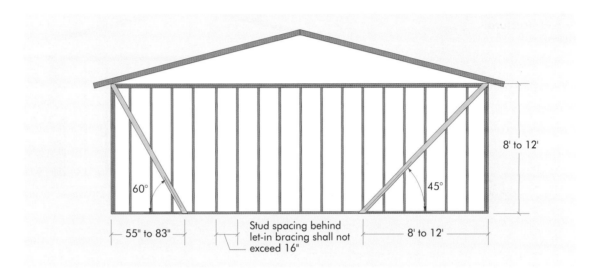

Method DWB (diagonal wood boards)

Method	Material	Minimum Thickness	Figure	Connection Criteria
DWB	Diagonal wood boards	3/4" (1" nominal) for maximum 24" stud spacing		2-8d (2-1/2" x 0.113") nails or 2 staples, 1-3/4" per stud

FIGURE 5.3

Method DWB (diagonal wood boards)

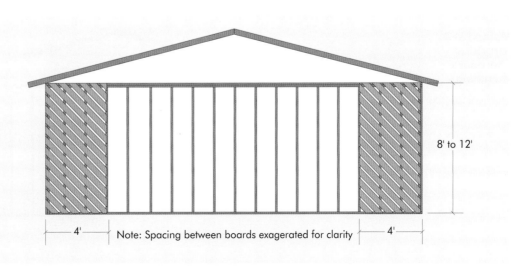

Method WSP (wood structural panel)

Method	Material	Minimum Thickness	Figure	Connection Criteria
WSP	Wood structural panel (See Section R604)	3/8"		Exterior sheathing: see IRC Table R602.3(3) Interior sheathing: see IRC Table R602.3(1)

FIGURE 5.4

Method WSP (wood structural panel)

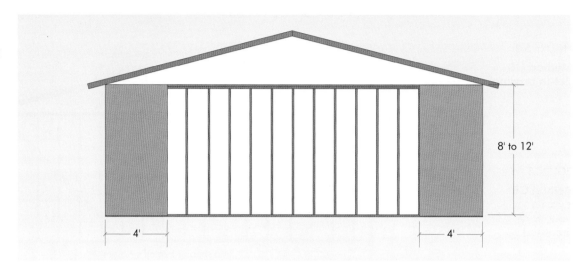

8' to 12'

4'   4'

Method SFB (structural fiberboard sheathing)

Note that the maximum stud spacing is 16 inches.

Method	Material	Minimum Thickness	Figure	Connection Criteria
SFB	Structural fiberboard sheathing	1/2" or 25/32" for maximum 16" stud spacing		1-1/2" galvanized roofing nails or 8d common (2-1/2" x 0.131") nails at 3" spacing (panel edges) at 6" spacing (intermediate supports)

FIGURE 5.5

Method SFB (structural fiberboard sheathing)

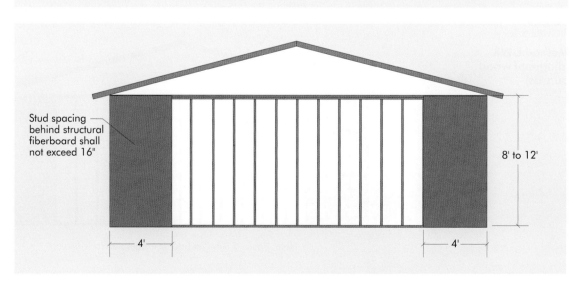

Stud spacing behind structural fiberboard shall not exceed 16"

8' to 12'

4'   4'

Method GB (gypsum board)

Note that fastening gypsum board per IRC Table R702.3.5 does not qualify as Method GB bracing. IRC Table R702.3.5 defines the attachment schedule when gypsum wall board is used as the required interior finish material for Methods LIB, WSP, SFB, PBS, PCP and HPS, or for gypsum sheathing applications unrelated to wall bracing.

Method	Material	Minimum Thickness	Figure	Connection Criteria
GB	Gypsum board	1/2"	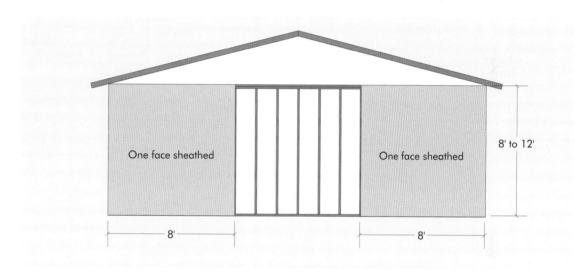	Nails and screws at 7" spacing at panel edges including top and bottom plates for all braced wall panel locations. For exterior sheathing nail or screw size, see IRC Table R602.3(1). For interior gypsum board nail or screw size, see IRC Table R702.3.5.

FIGURE 5.6a

Method GB (gypsum board) – one face sheathed

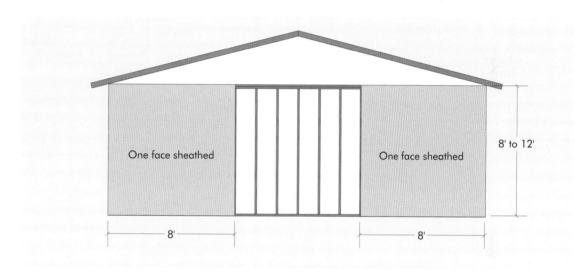

FIGURE 5.6b

Method GB (gypsum board) – two faces sheathed

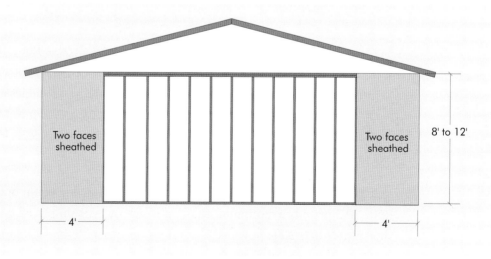

TABLE 5.2

Fastener requirements for Method GB (gypsum board) – gypsum sheathing

Excerpt from IRC Table R602.3(1) Fastening schedule for structural members

Item	Description of Building Materials	Description of Fastener[b]
	Other wall sheathing[h]	
36	1/2" gypsum sheathing[d]	1-1/2" galvanized roofing nail; staple galvanized, 1-1/2" long; 1-1/4" screws, Type W or S
37	5/8" gypsum sheathing[d]	1-3/4" galvanized roofing nail; staple galvanized, 1-5/8" long; 1-5/8" screws, Type W or S

b. Staples are 16-gage wire and have a minimum 7/16-inch on diameter crown width.
d. 4-foot-by-8-foot or 4-foot-by-9-foot.
h. Gypsum sheathing shall conform to ASTM C 1396 and shall be installed in accordance with GA 253. Fiberboard sheathing shall conform to ASTM C 208.

TABLE 5.3

Fastener requirements for Method GB (gypsum board) – interior gypsum board

Excerpt from IRC Table R702.3.5 with fastener spacing for braced wall panels

Thickness of Gypsum Board (inches)	Application	Orientation of Gypsum Board to Framing	Maximum Spacing of Framing Members (inches)	Maximum Spacing of Fasteners (For Bracing) (inches)		Size of Nails for Application to Wood Framing
				Nails	Screws[b]	
			Application without adhesive			
1/2	Wall	Either direction				13 gage, 1-3/8" long, 19/64" head; 0.098" diameter, 1-1/4" long, annular-ringed; 5d cooler nail, 0.086" diameter, 1-5/8" long, 15/64" head; or gypsum board nail, 0.086" diameter, 1-5/8" long, 9/32" head
	Wall	Either direction				
5/8	Wall	Either direction				13 gage, 1-5/8" long, 19/64" head; 0.098" diameter, 1-3/8" long, annular-ringed; 6d cooler nail, 0.092" diameter, 1-7/8" long, 1/4" head; or gypsum board nail, 0.0915" diameter, 1-7/8" long, 19/64" head
	Wall	Either direction				

b. Screws shall be in accordance with Section R702.3.6. Screws for attaching gypsum board to structural insulated panels shall penetrate the wood structural panel facing not less than 7/16 inch.

Note that the connection criteria for Method GB in IRC Table R602.10.2 specifies a nail and/or screw schedule of 7 inches at all panel edges; however, there is not sufficient space in the table to provide all of the permitted fastener types. Two additional tables addressing fastener requirements for gypsum sheathing and gypsum wall board are shown in **TABLES 5.2** and **5.3**. Gypsum sheathing, formulated for enhanced moisture and fire resistance, is intended for protected exterior wall applications.

The lengths indicated in **FIGURE 5.6a** are for single-sided gypsum bracing panels. IRC Section R602.10.3 permits a minimum length of 4 feet for gypsum placed on both faces of the wall, as shown in **FIGURE 5.6b**, provided that both sides meet Method GB fastening requirements.

Method PBS (particleboard sheathing)

Note the maximum stud spacing is 16 inches.

Method	Material	Minimum Thickness	Figure	Connection Criteria
PBS	Particleboard sheathing (See Section R605)	3/8" or 1/2" for maximum 16" stud spacing		1-1/2" galvanized roofing nails or 8d common (2-1/2" x 0.131") nails at 3" spacing (panel edges) at 6" spacing (intermediate supports)

FIGURE 5.7

Method PBS (particleboard sheathing)

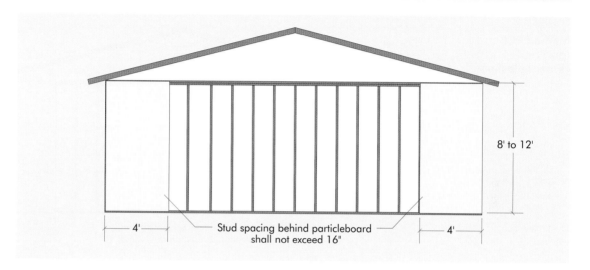

Note that in the 2009 IRC, the nail size and schedule has changed for particleboard when it is used for bracing. The traditional nailing schedule, per IRC Table R602.3(1) (6d common nails (2 inches x 0.113 inches) with a 6-inch and 12-inch schedule for panel edges and interior supports respectively), can still be used when the particleboard is <u>not used</u> for bracing.

The attachment schedule for particleboard sheathing when used as bracing has been increased to permit grouping of panel methods in IRC Tables R602.10.1.2(1) and R602.10.1.2(2).

Method PCP (portland cement plaster)

Note the maximum stud spacing is 16 inches.

Method	Material	Minimum Thickness	Figure	Connection Criteria
PCP	Portland cement plaster	See Section R703.6 for maximum 16" stud spacing		1-1/2", 11 gage, 7/16" head nails or 7/8", 16 gage staples at 6" spacing

FIGURE 5.8

Method PCP (portland cement plaster)

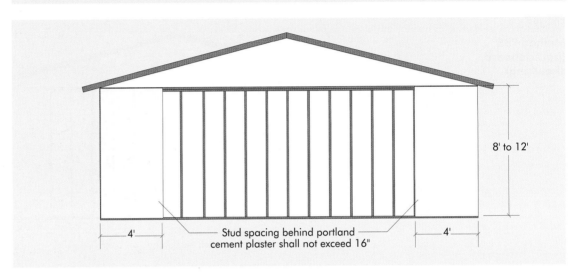

IRC Section R703.6 is referenced in the Minimum Thickness column. This section of the code includes additional thickness information and connection criteria.

Fastening requirements for bracing Method PCP:

R703.6 Exterior plaster. *Installation of these materials shall be in compliance with ASTM C 926 and ASTM C 1063 and the provisions of this code.*

R703.6.1 Lath. *All lath and lath attachment shall be of corrosion-resistant materials. Expanded metal or woven wire lath shall be attached with 1-1/2-inch-long (38 mm), 11-gage nails having a 7/16-inch (11.1 mm) head, or 7/8-inch-long (22.2 mm), 16-gage staples, spaced at no more than 6 inches (152 mm), or as otherwise approved.*

R703.6.2 Plaster. *Plastering with portland cement plaster shall not be less than three coats when applied to metal lath or wire lath and shall be not less than two coats when applied over masonry, concrete, pressure-preservative-treated wood or decay-resistant wood as specified in Section R317.1 or gypsum backing. If the plaster surface is completely covered by veneer or other facing material, or is completely concealed, plaster application need be only two coats, provided the total thickness is as set forth in IRC Table R702.1(1).*

Method HPS (hardboard panel siding)

Note the maximum stud spacing is 16 inches.

Method	Material	Minimum Thickness	Figure	Connection Criteria
HPS	Hardboard panel siding	7/16" for maximum 16" stud spacing		0.092" dia., 0.225" head nails with length to accommodate 1-1/2" penetration into studs at 4" spacing (panel edges), at 8" spacing (intermediate supports)

FIGURE 5.9

Method HPS (hardboard panel siding)

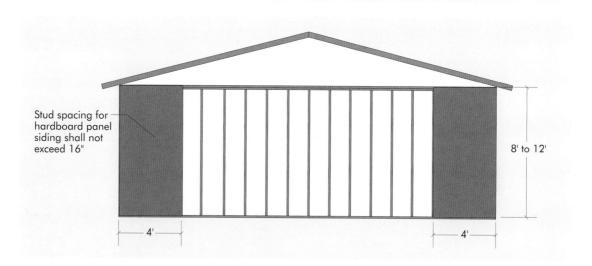

Stud spacing for hardboard panel siding shall not exceed 16"

4'   4'

8' to 12'

Note that in the 2009 IRC, the nail size and schedule for hardboard has been added to the table. This is not a change for 2009; rather, this information was formerly located in Footnotes m, n and o of the 2006 IRC Table R703.4.

Method ABW (alternate braced wall)

Method	Material	Minimum Thickness	Figure	Connection Criteria
ABW	Wood structural panel	3/8"		See Section R602.10.3.2

FIGURE 5.10

Method ABW (alternate braced wall)

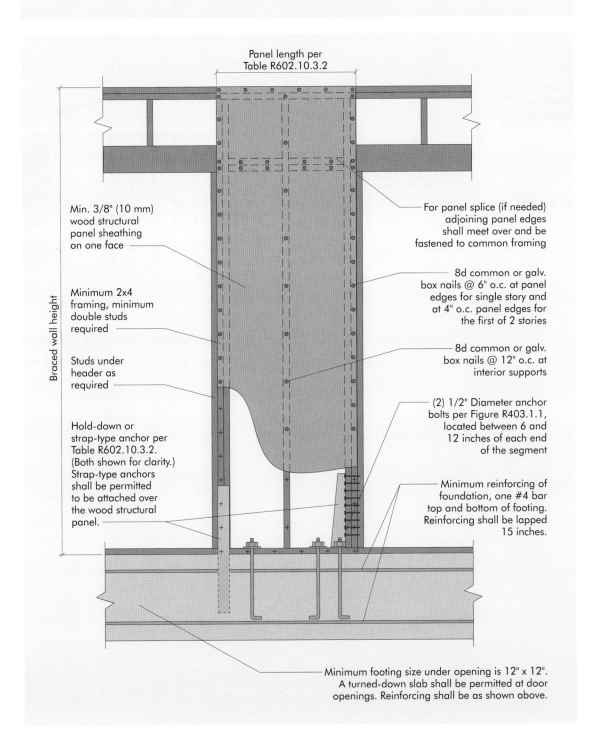

Panel length per Table R602.10.3.2

Min. 3/8" (10 mm) wood structural panel sheathing on one face

Minimum 2x4 framing, minimum double studs required

Studs under header as required

Hold-down or strap-type anchor per Table R602.10.3.2. (Both shown for clarity.) Strap-type anchors shall be permitted to be attached over the wood structural panel.

Braced wall height

For panel splice (if needed) adjoining panel edges shall meet over and be fastened to common framing

8d common or galv. box nails @ 6" o.c. at panel edges for single story and at 4" o.c. panel edges for the first of 2 stories

8d common or galv. box nails @ 12" o.c. at interior supports

(2) 1/2" Diameter anchor bolts per Figure R403.1.1, located between 6 and 12 inches of each end of the segment

Minimum reinforcing of foundation, one #4 bar top and bottom of footing. Reinforcing shall be lapped 15 inches.

Minimum footing size under opening is 12" x 12". A turned-down slab shall be permitted at door openings. Reinforcing shall be as shown above.

Method ABW (alternate braced wall) (IRC Section R602.10.3.2) has been in the IRC in one form or another since 2000 and was often referred to as the "32-inch alternate with hold downs." Method ABW panels are typically used when bracing is needed in a wall area that is not long enough to accommodate a 4-foot braced wall panel.

For the 2009 IRC, Method ABW was grouped with the intermittent wall bracing methods because it may be substituted on a one-for-one basis for a braced wall panel from any other bracing method, and, as such, it is used intermittently. A new figure (IRC Figure R602.10.3.2) was added to clarify the method, and reduce redundancy. Another change was the elimination of structural panels on both sides of the stud wall (double-sided) for the first-of-two-story applications: Instead of double-sided wood structural panels – which makes hold down and anchor bolt placement hard to inspect – the 2009 IRC requires nail spacing of 4 inches on center at the panel perimeter.

For purposes of computing the required length of bracing, a single Method ABW panel is considered to be equivalent to 4 feet of bracing, regardless of its actual length. Section R602.10.3.2 of the 2009 IRC is provided below, as is IRC Table R602.10.3.2 (**TABLE 5.4**).

> **R602.10.3.2 Method ABW: Alternate braced wall panels.** *Method ABW braced wall panels constructed in accordance with one of the following provisions shall be permitted to replace each 4 feet (1219 mm) of braced wall panel as required by Section R602.10.3. The maximum height and minimum length and hold down force of each panel shall be in accordance with IRC Table R602.10.3.2:*
>
> **1.** *In one-story buildings, each panel shall be installed in accordance with Figure R602.10.3.2. The hold-down device shall be installed in accordance with the manufacturer's recommendations. The panels shall be supported directly on a foundation or on floor framing supported directly on a foundation, which is continuous across the entire length of the braced wall line.*
>
> **2.** *In the first story of two-story buildings, each braced wall panel shall be in accordance with Item 1 above, except that the wood structural panel sheathing edge nailing spacing shall not exceed four inches (102 mm) on center.*

TABLE 5.4

Method ABW (alternate braced wall) – minimum hold down and length requirements

IRC Table R602.10.3.2

Seismic Design Category and Windspeed		Height of Braced Wall Panel				
		8 ft	9 ft	10 ft	11 ft	12 ft
SDC A, B and C Windspeed < 110 mph	Minimum Sheathed Length	2' 4"	2' 8"	2' 10"	3' 2"	3' 6"
	R602.10.3.2, Item 1 Hold Down Force (lbs)	1800	1800	1800	2000	2200
	R602.10.3.2, Item 2 Hold Down Force (lbs)	3000	3000	3000	3300	3600
SDC D$_0$, D$_1$ and D$_2$ Windspeed < 110 mph	Minimum Sheathed Length	2' 8"	2' 8"	2' 10"	NP[a]	NP[a]
	R602.10.3.2, Item 1 Hold Down Force (lbs)	1800	1800	1800	NP[a]	NP[a]
	R602.10.3.2, Item 2 Hold Down Force (lbs)	3000	3000	3000	NP[a]	NP[a]

For SI: 1 inch = 25.4 mm, 1 foot = 305 mm, 1 pound = 4.45 Newtons
a. NP = Not Permitted. Maximum height of 10 feet.

Method PFH (intermittent portal frame)

Method	Material	Minimum Thickness	Figure	Connection Criteria
PFH	Wood structural panel	3/8"		See Section R602.10.3.3

FIGURE 5.11

Method PFH (intermittent portal frame)

IRC Figure R602.10.3.3 Reformatted and modified to show foundation reinforcement requirements

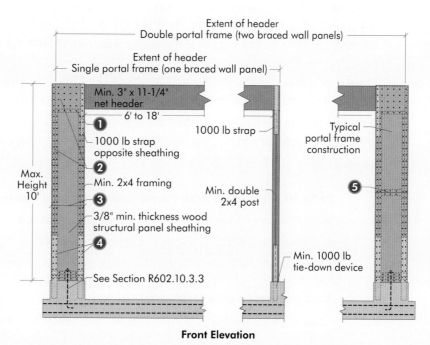

Extent of header
Double portal frame (two braced wall panels)

Extent of header
Single portal frame (one braced wall panel)

Min. 3" x 11-1/4" net header

6' to 18'

❶

1000 lb strap opposite sheathing

❷

Min. 2x4 framing

❸

3/8" min. thickness wood structural panel sheathing

❹

See Section R602.10.3.3

Max. Height 10'

1000 lb strap

Min. double 2x4 post

Min. 1000 lb tie-down device

Typical portal frame construction

❺

Front Elevation

❶ Fasten top plate to header with two rows of 16d sinker. Nails at 3 in. o.c. typ.

❷ Fasten sheathing to header with 8d common or galvanized box nails in 3 in. grid pattern as shown and 3 in. o.c. in all framing (studs, blocking, and sills) typ.

❸ Min. length = 16 in. for one story, min. length = 24 in. for use in the first of two story structures.

❹ Min. 4200 lb tie-down device (embedded into concrete and nailed into framing).

❺ For a panel splice (if needed), panel edges shall occur over and be nailed to common blocking, and occur within middle 24 in. of wall height. One row of 3 in. o.c. nailing is required in each panel edge.

Method PFH (intermittent portal frame) with hold downs has been in the IRC since 2006 (when it was called "alternate braced wall panel adjacent to a door or window opening" in the 2006 IRC Section R602.10.6.2). This method, also called the portal frame with hold downs in the text of the 2009 IRC, was developed primarily to maintain the traditional look of narrow wall segments on either side of garage door openings. Like Method ABW, Method PFH was included in the 2009 IRC as an intermittent bracing method because it may be substituted on a one-for-one basis for a braced wall panel from any other bracing method and, as such, it is used intermittently.

For purposes of computing the required length of bracing, each leg of the portal frame is considered to be equivalent to 4 feet of bracing. The portal frame can be built with one or two sides as required. This method can be used for wall heights up to 10 feet, or, if the amount of bracing in the wall line is increased by 20 percent, in walls up to 12 feet in height (when used for 12-foot walls, the clear opening height is limited to 10 feet minus the depth of the header). Note that Method PFH is <u>not</u> aspect ratio based: all single-story applications are a minimum of 16 inches in length and all first-of-two story applications are a minimum of 24 inches in length.

If a pony wall is to be used above Method PFH bracing, the pony wall details provided for Method CS-PF (continuous portal frame) (as shown in **FIGURE 5.22**) are applicable.

In the 2009 IRC, some text from theMethod PFH section was eliminated because it was redundant with the information presented in the figure. The 2009 abbreviated section is reproduced below. Note that the foundation must be reinforced as required by Method ABW (IRC Figure R602.10.3.2). **FIGURE 5.11** has been modified to show the updated foundation reinforcement requirements.

> **R602.10.3.3 Method PFH: Portal frame with hold downs.** *Method PFH braced wall panels constructed in accordance with one of the following provisions are also permitted to replace each 4 feet (1219 mm) of braced wall panel as required by Section R602.10.3 for use adjacent to a window or door opening with a full-length header:*
>
> **1.** *Each panel shall be fabricated in accordance with Figure R602.10.3.3. The wood structural panel sheathing shall extend up over the solid sawn or glued-laminated header and shall be nailed in accordance with Figure R602.10.3.3. A spacer, if used with built-up header, shall be placed on the side of the built-up beam opposite the wood structural panel sheathing. The header shall extend between the inside faces of the first full-length outer studs of each panel. One anchor bolt not less than 5/8-inch-diameter (16 mm) and installed in accordance with Section R403.1.6 shall be provided in the center of each sill plate. The hold-down devices shall be an embedded-strap type, installed in accordance with the manufacturer's recommendations. The panels shall be supported directly on a foundation, which is continuous across the entire length of the braced wall line. The foundation shall be reinforced as shown on Figure R602.10.3.2. This reinforcement shall be lapped not less than 15 inches (381 mm) with the reinforcement required in the continuous foundation located directly under the braced wall line.*
>
> **2.** *In the first story of two-story buildings, each wall panel shall be braced in accordance with Item 1 above, except that each panel shall have a length of not less than 24 inches (610 mm.)*

Method PFG (intermittent portal frame at garage door openings in Seismic Design Categories A, B and C)

Method	Material	Minimum Thickness	Figure	Connection Criteria
PFG	Wood structural panel	7/16"		See Section R602.10.3.4

FIGURE 5.12

Method PFG (intermittent portal frame at garage door openings in SDC A, B and C)

IRC Figure R602.10.3.4 Reformatted for clarity

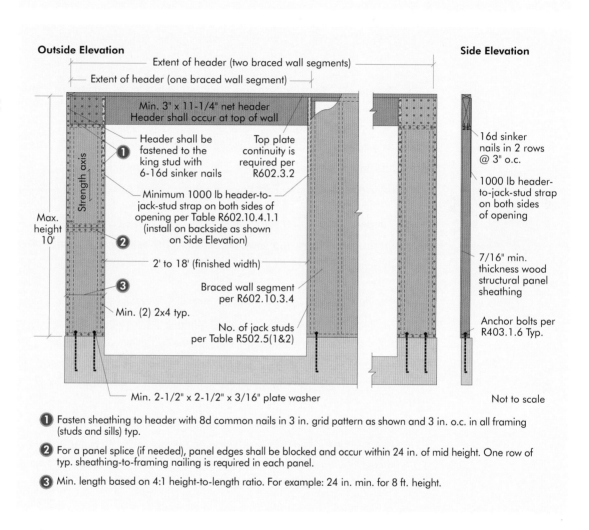

① Fasten sheathing to header with 8d common nails in 3 in. grid pattern as shown and 3 in. o.c. in all framing (studs and sills) typ.

② For a panel splice (if needed), panel edges shall be blocked and occur within 24 in. of mid height. One row of typ. sheathing-to-framing nailing is required in each panel.

③ Min. length based on 4:1 height-to-length ratio. For example: 24 in. min. for 8 ft. height.

Method PFG is new for the 2009 IRC and is a "light" variation of Method PFH (intermittent portal frame) that is limited to areas of low and moderate seismicity (SDC A, B and C). Like Method PFH, it is a portal frame bracing method, but it does not include hold downs and it is restricted to single-story and first-of-two-story applications. Method PFG is a 4:1 aspect ratio-based system with a 24-inch minimum length requirement (e.g., for an 8-foot wall, the minimum length is 96 inches/4 = 24 inches; for a 10-foot wall, the minimum length is 120 inches/4 = 30 inches). For the purposes of computing, its bracing length is equal to 1.5 times the length (horizontal dimension) of the vertical leg. Note the nailing at the sheathing-to-header overlap is the same as Method PFH, but without nailing the interior studs of the doubled 2x4s at either end of the vertical legs.

If a pony wall is used above Method PFG bracing, the pony wall details provided for Method CS-PF (continuous portal frame) (as shown in **FIGURE 5.22**) are applicable.

The provisions from IRC Section R602.10.3.4 are reproduced below:

> **R602.10.3.4 Method PFG portal frame at garage door openings in Seismic Design Categories A, B and C.** *Where supporting a roof or one story and a roof, alternate braced wall panels constructed in accordance with the following provisions are permitted on either side of garage door openings. For the purpose of calculating wall bracing amounts to satisfy the minimum requirements of Table R602.10.1.2(1), the length of the alternate braced wall panel shall be multiplied by a factor of 1.5.*
>
> 1. *Braced wall panel length shall be a minimum of 24 inches (610 mm) and braced wall panel height shall be a maximum of 10 feet (3048 mm).*
>
> 2. *Braced wall panel shall be sheathed on one face with a single layer of 7/16-inch-minimum (11 mm) thickness wood structural panel sheathing attached to framing with 8d common nails at 3 inches (76 mm) on center in accordance with Figure R602.10.3.4.*
>
> 3. *The wood structural panel sheathing shall extend up over the solid sawn or glued-laminated header and shall be nailed to the header at 3 inches (76 mm) on center grid in accordance with Figure R602.10.3.4.*
>
> 4. *The header shall consist of a minimum of two solid sawn 2x12s (51 by 305 mm) or a 3 inch x 11-1/4 inch (76 by 286 mm) glued-laminated header. The header shall extend between the inside faces of the first full-length outer studs of each panel in accordance with Figure R602.10.3.4. The clear span of the header between the inner studs of each panel shall be not less than 6 feet (1829 mm) and not more than 18 feet (5486 mm) in length.*
>
> 5. *A strap with an uplift capacity of not less than 1,000 pounds (4448 N) shall fasten the header to the side of the inner studs opposite the sheathing face. Where building is located in Wind Exposure Categories C or D, the strap uplift capacity shall be in accordance with Table R602.10.4.1.1.*
>
> 6. *A minimum of two bolts not less than 1/2-inch (12.7 mm) diameter shall be installed in accordance with Section R403.1.6. A 3/16-inch by 2-1/2-inch by 2-1/2-inch (4.8 by 63 by 63 mm) steel plate washer is installed between the bottom plate and the nut of each bolt.*
>
> 7. *Braced wall panel shall be installed directly on a foundation.*
>
> 8. *Where an alternate braced wall panel is located only on one side of the garage opening, the header shall be connected to a supporting jack stud on the opposite side of the garage opening with a metal strap with an uplift capacity of not less than 1,000 lbs (4448 N). Where that supporting jack stud is not part of a braced wall panel assembly, another 1,000 lbs (4448 N) strap shall be installed to attach the supporting jack stud to the foundation.*

Length requirements and adjustments for intermittent sheathing bracing methods

The 2009 IRC is more specific in detailing minimum panel lengths (the measured distance of the braced wall panel in the direction of the wall line) in IRC Section R602.10.3 (excerpted below) then in the past. This is due to an increase in the number of bracing types as well as a desire to address the ambiguity of the 2006 provisions.

This section requires that minimum braced panel length be at least 48 inches, except when bracing Method GB (gypsum board) is used: in such cases, the single-sided panel length should be at least 96 inches and 48 inches if double-sided. This section clarifies that, for panel-type bracing, the actual length of the panel is used to compute the length of bracing required. Method GB is again the exception: 96-inch single-sided panels are equal to 48 inches of bracing, while 48-inch double-sided panels are equal to 48 inches of bracing.

An exception is provided for continuous bracing methods because, in some circumstances, they permit bracing sections that are as little as 24 inches long. Like traditional panel bracing, the actual length of the panels is used to compute the amount of bracing required.

Also excepted are Method ABW (alternate braced wall), Method PFH (intermittent portal frame), and Method PFG (intermittent portal frame at garage door openings in Seismic Design Categories A, B and C). These are not strictly panel-type methods and their bracing capacity is relative to more than panel length. Lengths of Method ABW, PFH and PFG panels should be in accordance with IRC Sections R602.10.3.2 (equal to 4 feet of bracing), R602.10.3.3 (equal to 4 feet of bracing), and R602.10.3.4 (equal to 1.5 times the length of the vertical leg measured along the length of the braced wall line), respectively.

IRC Section R602.10.3, Exception 4 permits reductions in the minimum panel length for specific panel-type bracing methods (DWB, WSP, SFB, PCP, PBS and HPS) in SDC A, B and C, as long as a length penalty is applied to the actual length of the wall panel (see IRC Table R602.10.3, provided in **TABLE 5.5**). For example, a 36-inch panel can be used in an 8-foot wall instead of a full 48-inch panel, but it will only count as 27 inches towards the required length of bracing.

> **R602.10.3 Minimum length of braced panels.** *For Methods DWB, WSP, SFB, PBS, PCP and HPS, each braced wall panel shall be at least 48 inches (1219 mm) in length, covering a minimum of three stud spaces where studs are spaced 16 inches (406 mm) on center and covering a minimum of two stud spaces where studs are spaced 24 inches (610 mm) on center. For Method GB, each braced wall panel and shall be at least 96 inches (2438 mm) in length where applied to one face of a braced wall panel and at least 48 inches (1219 mm) where applied to both faces. For Methods DWB, WSP, SFB, PBS, PCP and HPS, for purposes of computing the length of panel bracing required in Table R602.10.1.2(1) and R602.10.1.2(2), the effective length of the braced wall panel shall be equal to the actual length of the panel. When Method GB panels are applied to only one face of a braced wall panel, bracing lengths required in Table R602.10.1.2(1) and R602.10.1.2(2) for Method GB shall be doubled.*

Exceptions:

1. *Lengths of braced wall panels for continuous sheathing methods shall be in accordance with Table R602.10.4.2.*

2. *Lengths of Method ABW panels shall be in accordance with Section R602.10.3.2.*

3. *Length of Methods PFH and PFG shall be in accordance with Section R602.10.3.3 and R602.10.3.4, respectively.*

4. *For Methods DWB, WSP, SFB, PBS, PCP and HPS in Seismic Design Categories A, B, and C: panels between 36 inches (914 mm) and 48 inches (1219 mm) in length shall be permitted, to count towards the required length of bracing in Tables R602.10.1.2(1) and R602.10.1.2(2), and the effective contribution shall comply with Table R602.10.3.*

TABLE 5.5

Effective lengths for braced wall panels less than 48 inches in actual length (bracing methods DWB, WSP, SFB, PBS, PCP and HPS[a]

IRC Table R602.10.3 Length penalties for panels less than 48 inches long

Actual Length of Braced Wall Panel (inches)	Effective Length of Braced Wall Panel (inches)		
	8-foot Wall Height	9-foot Wall Height	10-foot Wall Height
48	48	48	48
42	36	36	N/A
36	27	N/A	N/A

For SI: 1 inch = 25.4mm, 1 foot = 304.5 mm.
a. Interpolation shall be permitted.

IRC Table R602.10.3.1 is a new addition to the IRC (shown in **TABLE 5.6**). This table modifies the minimum bracing length requirements in IRC Section R602.10 3 when braced wall lines are greater than 10 feet in height.

In the 2009 IRC, Section R301.3.1, Exception allows interpolation of the bracing requirement to increase with increased height above 10 feet. **TABLE 5.6** contains the required bracing lengths.

TABLE 5.6

Minimum length requirements for braced wall panels

IRC Table R602.10.3.1

Seismic Design Category And Wind Speed	Bracing Method	Height Of Braced Wall Panel				
		8 ft	9 ft	10 ft	11 ft	12 ft
SDC A, B, C, D_0, D_1 and D_2 Wind speed < 110 mph	DWB, WSP, SFB, PBS, PCP, HPS and Method GB when double-sided	4' 0"	4' 0"	4' 0"	4' 5"	4' 10"
	Method GB, single-sided	8' 0"	8' 0"	8' 0"	8' 10"	9' 8"

For SI: l inch = 25.4mm, 1 foot = 305 mm

Note that this aspect ratio adjustment is separate from the required bracing adjustment in IRC Section R301.3, Item 1, Exception. However, any increase in length as a result of the aspect ratio adjustment is permitted to be counted towards the amount of bracing required. In many cases, when a number of minimum-length panels are used for bracing, the aspect ratio adjustment will completely account for the length increase. This section of the 2009 IRC is provided below:

> **R602.10.3.1 Adjustment of length of braced panels.** *When story height (H), measured in feet, exceeds 10 feet (3048 mm), in accordance with Section R301.3, the minimum length of braced wall panels specified in Section R602.10.3 shall be increased by a factor H/10. See Table R602.10.3.1. Interpolation is permitted.*

Description of continuous sheathing bracing methods

Method CS-WSP (continuous wood structural panel sheathing), originally introduced in the 2000 IRC, has not changed much for the 2009 IRC. Method CS-SFB (continuous structural fiberboard sheathing) is new to the 2009 edition. Another change is that some specific bracing provisions, previously included as footnotes and only permitted when using Method CS-WSP, have been redefined as bracing methods. These new additions are incorporated into the new IRC Table R602.10.4.1 (reproduced in **TABLE 5.7**).

The two basic continuous sheathing bracing methods, Method CS-WSP (continuous wood structural panel sheathing) and Method CS-SFB (continuous structural fiberboard sheathing) (described in IRC Sections R602.10.4 and R602.10.5, respectively) require all braced wall lines, including areas above and below openings and gable ends, to be fully sheathed with a minimum of 3/8-inch structural panel sheathing (as shown in **FIGURE 5.13**) or 1/2-inch structural fiberboard sheathing.

FIGURE 5.13

Example of continuously sheathed braced walls

In addition to continuously sheathing each wall, all corners must be constructed in accordance with IRC Figure R602.10.4.4(1), as shown in **FIGURE 5.14**.

FIGURE 5.14

Required exterior corner framing attachment details for continuous sheathing

IRC Figure R602.10.4.4.1

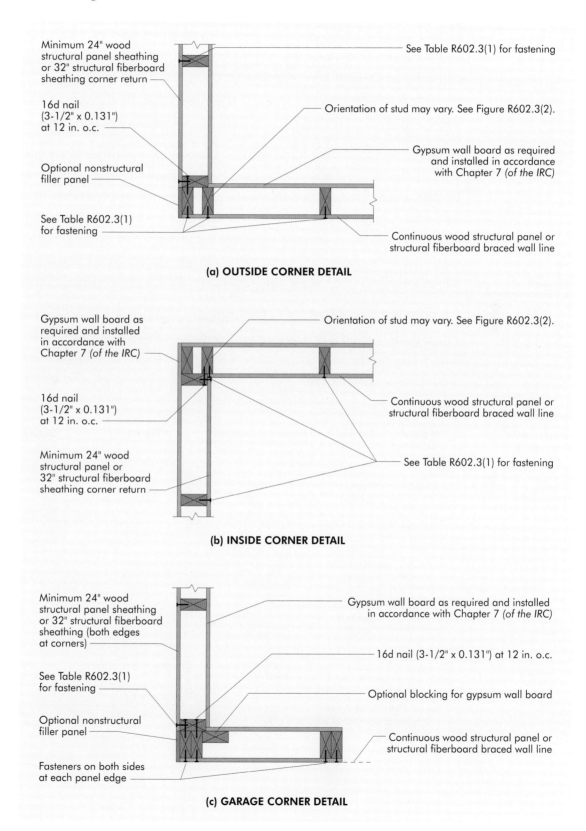

(a) OUTSIDE CORNER DETAIL

(b) INSIDE CORNER DETAIL

(c) GARAGE CORNER DETAIL

The purpose of the corner attachment for the continuous sheathing bracing methods is to connect the intersecting walls together to create a stronger, box-like structure that will perform better during high wind or seismic events. The corner detail requires a minimum of a single row of 16d nails at 12 inches on center. A double row of 16d nails at 24 inches on center, framing member orientation permitting, is considered equivalent. It is important to note that the intent of IRC Figure R602.10.4.4(1) is to provide the specified amount of nailing between the two studs, each on adjacent walls, to which the adjacent wall sheathing is attached. Additional information about corner requirements and options are provided in **CHAPTER 7**.

Because of the added strength and stiffness of wood structural panels when applied to all exterior surfaces (that are not window or door openings), narrow-width bracing panels may be used, allowing for additional architectural latitude. Panel lengths as narrow as 24 inches (4:1 aspect ratio) can be used adjacent to garage door openings with very light roof loads (IRC Table R602.10.4.1, Footnote b, Method CS-G (wood structural panel adjacent to garage door openings and supporting roof loads only)). Due to the required 4:1 aspect ratio, taller walls require longer braced panels. In addition, the minimum length of the bracing panel depends on the size of the opening next to the bracing panel.

The 2009 IRC also permits the use of site-built portal frames without hold downs adjacent to garage doors with an aspect ratio of 6:1 (Method CS-PF (continuous portal frame)). These portal frames may now be used on any story, over elevated floors; however, no more than four may be used in any single wall line.

TABLE 5.7

Continuous sheathing methods

IRC Table R602.10.4.1

Method	Material	Minimum Thickness	Figure	Connection Criteria
CS-WSP	Wood structural panel	3/8"		6d common (2" x 0.113") nails at 6" spacing (panel edges) and at 12" spacing (intermediate supports) or 16 ga. x 1-3/4 staples: at 3" spacing (panel edges) and 6" spacing (intermediate supports)
CS-G	Wood structural panel adjacent to garage openings and supporting roof load only[a,b]	3/8"		See Method CS-WSP
CS-PF	Continuous portal frame	See Section R602.10.4.1.1		See Section R602.10.4.1.1

For SI: 1 inch = 25.4 mm
a. Applies to one wall of a garage only.
b. Roof covering dead loads shall be 3 psf or less.

Method CS-WSP (continuous wood structural panel sheathing)

Method	Material	Minimum Thickness	Figure	Connection Criteria
CS-WSP	Wood structural panel	3/8"		6d common (2" x 0.113") nails at 6" spacing (panel edges) and at 12" spacing (intermediate supports) or16 ga. x 1-3/4 staples: at 3" spacing (panel edges) and 6" spacing (intermediate supports)

FIGURE 5.15

Method CS-WSP (continuous wood structural panel sheathing)

IRC Section R602.10.4

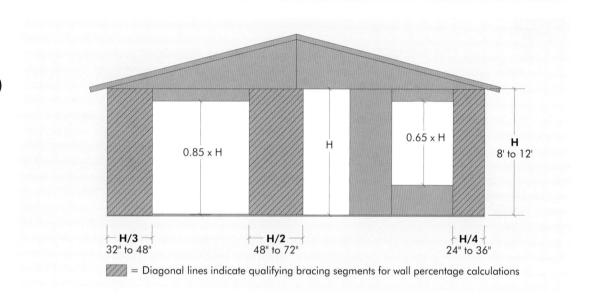

= Diagonal lines indicate qualifying bracing segments for wall percentage calculations

This portion of the 2009 IRC is taken from the previous edition, but includes a number of changes. The most significant change is that within SDC A, B and C, or in regions where the basic wind speed is less than or equal to 100 mph, there is no requirement for Method CS-WSP (continuous wood structural panel sheathing) to be used on *all* exterior walls of a story using continuous sheathing. In the 2009 IRC, Method CS-WSP is permitted to be used on a wall-by-wall, story-by-story basis, with other walls or stories of the structure utilizing any approved bracing method in the IRC.

In SDC D_0, D_1 and D_2, all exterior braced wall lines on the same story must be continuously sheathed before Method CS-WSP can be used.

Mixing bracing methods is covered in Section R602.10.1.1 of the 2009 IRC and will be addressed later in this chapter, but the only limitation that is specific to the continuous bracing methods is that no other methods may be used in a continuously sheathed wall line with one exception:

- Method CS-G (wood structural panel adjacent to garage door opening and supporting roof loads only) and Method CS-PF (continuous portal frame) may be used in a Method CS-WSP (continuous wood structural panel sheathing) wall line, as these methods are predicated on their use in a continuously sheathed wood structural panel wall line (IRC Section R602.10.4.1).

Continuously sheathed wall lines are permitted on a wall line by wall line basis in the 2009 IRC (except in SDC D_0, D_1 and D_2) and a number of requirements for corner details are provided; however, Method CS-WSP can be placed away from corners, as is permitted for intermittent methods. A number of figures have been added to the 2009 IRC to clarify the end attachment requirements of the continuous sheathing provisions (see **FIGURES 5.16** through **5.19** for illustrations of corner details). Additional information regarding corner details is provided in **CHAPTER 7**.

FIGURE 5.16

Continuous sheathing methods

IRC Figure R602.10.4.4(2) Braced wall line with continuous sheathing with corner return panel per IRC Section R602.10.4.4 or R602.10.5. Note that return corner must be 32 inches when using IRC Section R602.10.5, Method CS-SFB.

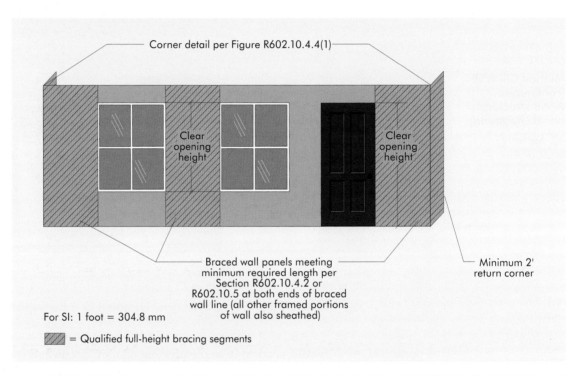

FIGURE 5.17

Continuous sheathing methods

IRC Figure R602.10.4.4(3) Braced wall line with continuous sheathing, with hold down but without corner return panel per IRC Section R602.10.4.4 or R602.10.5

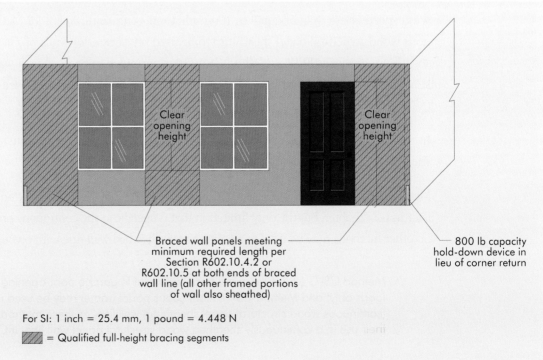

FIGURE 5.18

Continuous sheathing methods

IRC Figure R602.10.4.4(4) Braced wall line with continuous sheathing - first braced wall panel away from end of wall line without hold down per IRC Section R602.10.4.4 or R602.10.5. Note that return corner must be 32 inches when using IRC Section R602.10.5, Method CS-SFB.

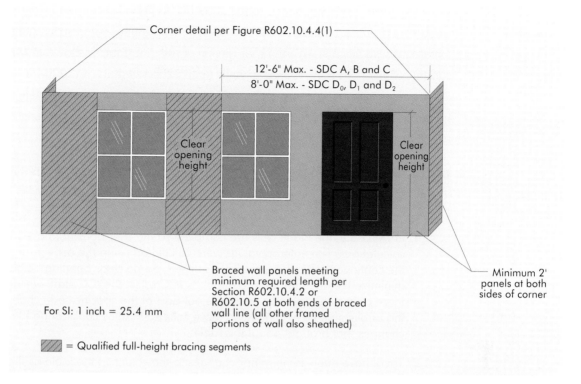

Corner detail per Figure R602.10.4.4(1)

12'-6" Max. - SDC A, B and C
8'-0" Max. - SDC D$_0$, D$_1$ and D$_2$

Clear opening height

Clear opening height

Minimum 2' panels at both sides of corner

Braced wall panels meeting minimum required length per Section R602.10.4.2 or R602.10.5 at both ends of braced wall line (all other framed portions of wall also sheathed)

For SI: 1 inch = 25.4 mm

▨ = Qualified full-height bracing segments

FIGURE 5.19

Continuous sheathing methods

IRC Figure R602.10.4.4(5) Braced wall line with continuous sheathing – first braced wall panel away from end of wall line with hold down per IRC Section R602.10.4.4 or R602.10.5

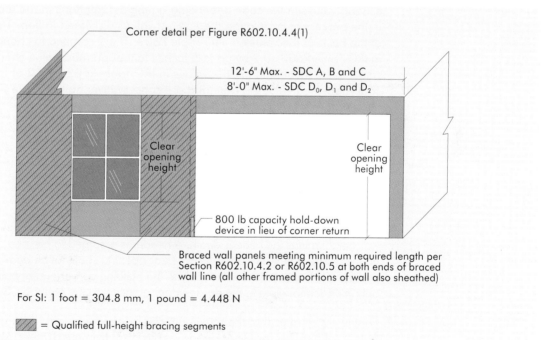

Corner detail per Figure R602.10.4.4(1)

12'-6" Max. - SDC A, B and C
8'-0" Max. - SDC D$_0$, D$_1$ and D$_2$

Clear opening height

Clear opening height

800 lb capacity hold-down device in lieu of corner return

Braced wall panels meeting minimum required length per Section R602.10.4.2 or R602.10.5 at both ends of braced wall line (all other framed portions of wall also sheathed)

For SI: 1 foot = 304.8 mm, 1 pound = 4.448 N

▨ = Qualified full-height bracing segments

IRC Table R602.10.4.2 was added to the 2009 IRC to provide minimum length requirements for wood structural panels used in Method CS-WSP (continuous wood structural panel sheathing) for various wall heights. An enhanced version of this IRC Table is shown in **TABLE 5.8**. Note that continuous methods may be used in walls up to 12 feet high in the 2009 IRC (a clarification of the 2006 code that did not address the issue). Also note that minimum panel lengths are aspect-ratio linked: as the wall height increases, so does the minimum bracing panel length. This is necessary to ensure that, as the wall gets taller, it remains sufficiently stiff.

TABLE 5.8 is an enhanced version of IRC Table R602.10.4.2, differing from the original in two key ways:

1. In **TABLE 5.8**, the blank cells are complete so the user does not have to do the calculations. The entered values were calculated using the aspect ratios upon which the continuous sheathing method is based. Some clear opening heights were added to minimize the need for interpolation. By ICC rules, the ICC staff was unable to complete these cells in this table as they were not part of the specific code change. The authors of this publication have completed these cells editorially, considering the intent of the code changes that brought about this table.

2. There have been some minor changes to the values in the table as they appear in IRC Table R602.10.4.2 in the 2009 IRC. (There are a number of ways to interpolate the existing values in the code depending on the assumptions made regarding those values. In the process of calculating the entered values discussed in Item 1 above, a closer look at the basis of the values in IRC Table R602.10.4.2 resulted in changes to some values in IRC Table R602.10.2 by up to 5 inches (although most value changes were just by a single inch). In the interim, the values in IRC Table R602.10.4.2 of the 2009 IRC are the code-approved values to date and are conservative; however, the values in **TABLE 5.8** are correct from an engineering perspective.)

Note also that the wall height values for Method CS-G (wood structural panel adjacent to garage door openings and supporting roof loads only) and Method CS-PF (continuous portal frame) for 11 and 12 feet are included and footnoted:

1. Footnote b is added to clarify that Method CS-G assumes clear opening height equal to the wall height minus the header depth. A header is assumed to be used over the garage door opening. The statement "a full-height clear opening shall not be permitted adjacent to a Method CS-G panel" does not apply to the building corner/end of building. This statement is only intended to relate to the garage door adjacent to the Method CS-G panel.

2. Footnote c clarifies that the maximum clear opening height (distance from grade to bottom of header) for the Method CS-PF portal frame is 10 feet. Method CS-PF may be used in a 12 foot wall with the difference in height made up for by the use of a pony wall over the portal frame. IRC Figure R602.10.4.1.1 and IRC Table R602.10.4.1.1 in the 2009 IRC provide information as to how to properly construct pony walls for various wind speeds. Corrected IRC Figure R602.10.4.1.1 is shown in errata to the 2009 IRC. This can be viewed at www.iccsafe.org/cs/codes/errata.html. Maximum pony wall heights are given in IRC Table R602.10.4.1.1.

TABLE 5.8

Length requirements for braced wall panels with continuous sheating[a]

Enhanced and corrected IRC Table R602.10.4.2

Method	Adjacent Clear Opening Height	Wall Height				
		8'	9'	10'	11'	12'
CS-WSP	64"	24"	27"	30"	33"	36"
	68"	26"	27"	30"	33"	36"
	72"	27"	27"	30"	33"	36"
	76"	30"	29"	30"	33"	36"
	80"	32"	30"	30"	33"	36"
	84"	35"	32"	32"	33"	36"
	88"	38"	35"	33"	33"	36"
	92"	43"	37"	35"	35"	36"
	96"	48"	41"	38"	36"	36"
	100"		44"	40"	38"	38
	104"		49"	43"	40"	39
	108"		54"	46"	43"	41
	112"			50"	45"	43
	116"			55"	48"	45
	120"			60"	52"	48
	124"				56"	51
	128"				61"	54
	132"				66"	58
	136"					62
	140"					66
	144"					72"
CS-G	See note b	24"	27"	30"	33"	36"
CS-PF	≤ 120"	16"	18"	20"	22"[c]	24"[c]

For SI: 1 inch = 25.4 mm, 1 foot = 305 mm

a. Interpolation shall be permitted.

b. Garage openings adjacent to a method CS-G panel shall be provided with a header in accordance with IRC Table R502.5(1). A full-height clear opening shall not be permitted adjacent to a method CS-G panel.

c. Maximum opening height shall be in accordance with IRC Figure R602.10.4.1.1.

The sections of the 2009 IRC covering Method CS-WSP (continuous wood structural panel sheathing) are reproduced below.

> **R602.10.4 Continuous sheathing.** *Braced wall lines with continuous sheathing shall be constructed in accordance with this section. All braced wall lines along exterior walls on the same story shall be continuously sheathed.*
>
> **Exception:** *Within Seismic Design Categories A, B, and C or in regions where the basic wind speed is less than or equal to 100 miles per hour (45 m/s), other bracing methods prescribed by this code shall be permitted on other braced wall lines on the same story level or on any braced wall line on different story levels of the building.*

R602.10.4.1 Continuous sheathing braced wall panels. *Continuous sheathing methods require structural panel sheathing to be used on all sheathable surfaces on one side of a braced wall line including areas above and below openings and gable end walls. Braced wall panels shall be constructed in accordance with one of the methods listed in Table R602.10.4.1. Different bracing methods, other than those listed in Table R602.10.4.1, shall not be permitted along a braced wall line with continuous sheathing.*

R602.10.4.2 Length of braced wall panels with continuous sheathing. *Braced wall panels along a braced wall line with continuous sheathing shall be full-height with a length based on the adjacent clear opening height in accordance with Table R602.10.4.2 and Figure R602.10.4.2. Within a braced wall line when a panel has an opening on either side of differing heights, the taller opening height shall be used to determine the panel length from Table R602.10.4.2. For Method CS-PF, wall height shall be measured from the top of the header to the bottom of the bottom plate as shown in Figure R602.10.4.1.1.*

R602.10.4.3 Length of bracing for continuous sheathing. *Braced wall lines with continuous sheathing shall be provided with braced wall panels in the length required in Table R602.10.1.2(1) and R602.10.1.2(2). Only those full-height braced wall panels complying with the length requirements of Table R602.10.4.2 shall be permitted to contribute towards the minimum required length of bracing.*

R602.10.4.4 Continuously sheathed braced wall panel location and corner construction. *For all continuous sheathing methods, full-height braced wall panels complying with the length requirements of Table R602.10.4.2 shall be located at each end of a braced wall line with continuous sheathing and at least every 25 feet (7620 mm) on center. A minimum 24-inch (610 mm) wood structural panel corner return shall be provided at both ends of a braced wall line with continuous sheathing in accordance with Figures R602.10.4.4(1) and R602.10.4.4(2). In lieu of the corner return, a hold-down device with a minimum uplift design value of 800 lb (3560 N) shall be fastened to the corner stud and to the foundation or framing below in accordance with Figure R602.10.4.4(3). (Corner details are covered in* **CHAPTER 7**.*)*

Exception: *The first braced wall panel shall be permitted to begin 12.5 feet (3810 mm) from each end of the braced wall line in Seismic Design Categories A, B, and C and 8 feet (2438 mm) in Seismic Design Categories D_0, D_1 and D_2 provided one of the following is satisfied:*

1. *A minimum 24 inch long (610 mm), full-height wood structural panel is provided at both sides of a corner constructed in accordance with Figure R602.10.4.4(1) at the braced wall line ends in accordance with Figure R602.10.4.4(4), or*

2. *The braced wall panel closest to the corner shall have a hold-down device with a minimum uplift design value of 800 lb (3650 N) fastened to the stud at the edge of the braced wall panel closest to the corner and to the foundation or framing below in accordance with Figure R602.10.4.4(5).*

Additional information on corner and end distance requirements can be found in **CHAPTER 7**.

Method CS-G (wood structural panel adjacent to garage door openings and supporting roof loads only)

Method	Material	Minimum Thickness	Figure	Connection Criteria
CS-G	Wood structural panel adjacent to garage openings and supporting roof load only[a,b]	3/8"		See Method CS-WSP

For SI: 1 inch = 25.4 mm
a. Applies to one wall of a garage only.
b. Roof covering dead loads shall be 3 psf or less.

FIGURE 5.20

Method CS-G (wood structural panel adjacent to garage door openings and supporting roof loads only)

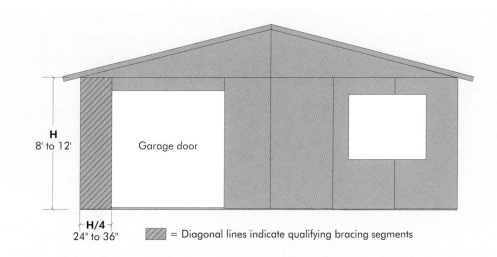

H 8' to 12'

Garage door

H/4 24" to 36"

▨ = Diagonal lines indicate qualifying bracing segments

Method CS-G (wood structural panel adjacent to garage door openings and supporting roof loads only) is a narrow-length panel that can be used with Method CS-WSP (continuous wood structural panel sheathing). In the 2006 IRC, this method is mentioned only in Footnote b of the 2006 IRC Table R602.10.5. This method permits the use of a panel length as short as 24 inches for an 8-foot wall, but because the length of the wall is linked to an aspect ratio (height/length = 4/1), the segment is required to get longer as the wall gets taller.

In IRC Table R602.10.4.2 of the 2009 edition, the aspect ratio description has been dropped and minimum lengths are used, based on the 4:1 aspect ratio. (See *TABLE 5.8*. Note that this table provides the length requirement for 11- and 12-foot high walls.)

Pay particular attention to the two footnotes to IRC Table R602.10.4.1 (*TABLE 5.7*). The first does not permit the use of 4:1 aspect ratio narrow wall panels on more than one wall of a garage. In other words, the braced panel may only be used around the garage door opening. If garage door openings occur in two walls of the garage, only one wall may use the braced panel. This should not be interpreted to mean that narrow wall segments cannot be used on both sides of a garage door within a single wall; rather, that is their designed use.

The second footnote restricts 4:1 aspect ratio narrow wall segments to supporting roofs only, and further restricts the dead load of the roof covering to 3 psf. This minimizes the seismic weight (mass) of the roof in a seismic event by limiting the roof covering to relatively light roofing materials. As such, this restriction is only applicable to structures that are not exempt from the seismic requirements of the IRC.

Method CS-G can only be used with Method CS-WSP. For purposes of determining length of bracing in a wall line with Method CS-G, use the actual length of the narrow braced wall panel(s). Of course, the required length of bracing must be met for any wall in the structure: this provision simply allows the use of smaller segments to make up the required length of bracing.

Method CS-PF (continuous portal frame)

Method	Material	Minimum Thickness	Figure	Connection Criteria
CS-PF	Wood structural panel	3/8"		See Section R602.10.4.1.1

FIGURE 5.21

Method CS-PF (continuous portal frame)

IRC Section R602.10.4.1.1

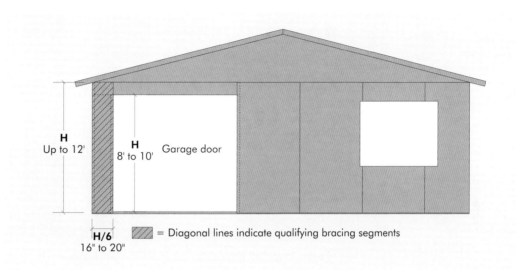

H Up to 12'

H 8' to 10'   Garage door

H/6 16" to 20"

= Diagonal lines indicate qualifying bracing segments

In the 2006 IRC, this method was specified in Footnote c of the 2006 IRC Table R602.10.5 and it was known in the industry as "portal frame without hold downs". In the 2009 IRC, the scope of this method, now referred to as Method CS-PF (continuous portal frame), has been expanded for use on any floor and in any wind zone or SDC covered by the IRC. The expansion to any floor comes as the result of the development of attachment details for raised wood floors in addition to foundation details. For purposes of determining length of bracing, the length of the vertical leg of the portal frame is used as the bracing length for the element. Remember that, for all bracing methods, wall length is measured along the wall line in the horizontal direction. Note that the annotations in *FIGURE 5.22* provide guidance on the minimum length of Method CS-PF:

Min. length based on 6:1 height-to-length ratio: for example: 16" min. for 8' height.

Minimum length information is also included in IRC Table R602.10.4.2 (*TABLE 5.8*).

FIGURE 5.22

Method CS-PF – continuous portal frame panel construction

IRC Figure R602.10.4.1.1 (2009 Errata) Reformatted for clarity.

Details of portal frame without hold downs.

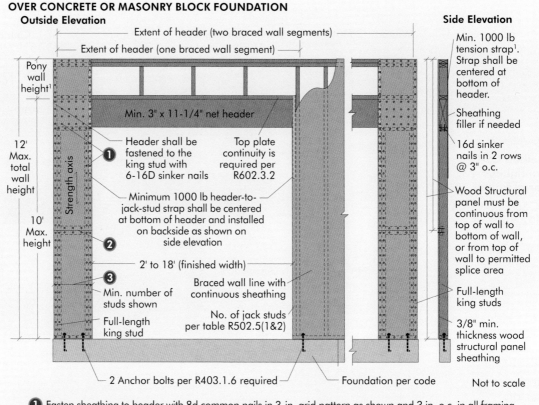

OVER CONCRETE OR MASONRY BLOCK FOUNDATION
Outside Elevation

Extent of header (two braced wall segments)

Extent of header (one braced wall segment)

Pony wall height[1]

12' Max. total wall height

10' Max. height

Strength axis

Min. 3" x 11-1/4" net header

① Header shall be fastened to the king stud with 6-16D sinker nails

Top plate continuity is required per R602.3.2

Minimum 1000 lb header-to-jack-stud strap shall be centered at bottom of header and installed on backside as shown on side elevation

②

2' to 18' (finished width)

③ Min. number of studs shown

Full-length king stud

Braced wall line with continuous sheathing

No. of jack studs per table R502.5(1&2)

2 Anchor bolts per R403.1.6 required

Foundation per code

Side Elevation

Min. 1000 lb tension strap[1]. Strap shall be centered at bottom of header.

Sheathing filler if needed

16d sinker nails in 2 rows @ 3" o.c.

Wood Structural panel must be continuous from top of wall to bottom of wall, or from top of wall to permitted splice area

Full-length king studs

3/8" min. thickness wood structural panel sheathing

Not to scale

① Fasten sheathing to header with 8d common nails in 3-in. grid pattern as shown and 3 in. o.c. in all framing (studs and sills) typ.

② For a panel splice (if needed), panel edges shall occur over and be nailed to common blocking and occur within the middle 24 in. of wall height. One row of 3 in. o.c. nailing is required at each panel edge.

③ Min. length based on 6:1 height-to-width ratio. For example: 16 in. min. for 8 ft. height.

[1]Per table R602.10.4.1.1

OVER RAISED WOOD FLOORS OR SECOND FLOOR – FRAMING ANCHOR OPTION

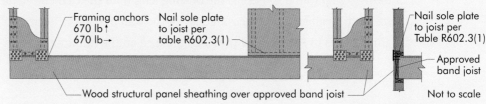

Framing anchors 670 lb ↑ 670 lb →

Nail sole plate to joist per table R602.3(1)

Nail sole plate to joist per Table R602.3(1)

Approved band joist

Wood structural panel sheathing over approved band joist

Not to scale

OVER RAISED WOOD FLOORS OR SECOND FLOOR – WOOD STRUCTURAL PANEL OVERLAP OPTION

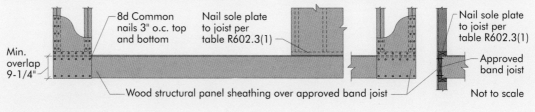

8d Common nails 3" o.c. top and bottom

Nail sole plate to joist per table R602.3(1)

Nail sole plate to joist per table R602.3(1)

Min. overlap 9-1/4"

Approved band joist

Wood structural panel sheathing over approved band joist

Not to scale

Method CS-PF can only be used with Method CS-WSP (continuous wood structural panel sheathing).

Also new for 2009 are provisions for the use of a pony wall directly over the portal frame, which is often, but not limited to, a means of elevating the second story of the structure over the garage in a home with a split-level entry. These provisions are meant to address a problem in which a structural hinge is created over a door or window header that can result in the header bulging in or out due to wind loads blowing directly against the wall, or even differential moisture conditions. In severe cases, it can lead to structural problems. Often in garages, these areas are braced back to the ceiling with framing to prevent such an occurrence.

In a similar manner, a new requirement in Section R602.3 of the 2009 IRC requires studs to be continuous from the sole plate to the top plate to resist loads perpendicular to the plane of the wall.

When "braced back" detailing is not desired, the information provided in IRC Figure R602.10.4.1.1 (**FIGURE 5.22**) and in IRC Table R602.10.4.1.1 (**TABLE 5.9**) can be used to detail the portal frame to handle various design wind loads and exposures. While not specifically permitted in the IRC, from an engineering perspective, interpolation of the variables in the table is appropriate.

TABLE 5.9

Tension strap capacity required for resisting wind pressures perpendicular to 6:1 aspect ratio walls[a,b]

IRC Table R602.10.4.1.1

Minimum Wall Stud Framing Nominal Size and Grade	Maximum Pony Wall Height (ft)	Maximum Total Wall Height (ft)	Maximum Opening Width (ft)	Basic Wind Speed (mph)					
				85	90	100	85	90	100
				Exposure B			Exposure C		
				Tension strap capacity required (lbf)[a,b]					
2x4 No. 2 Grade	0	10	18	1000	1000	1000	1000	1000	1000
	1	10	9	1000	1000	1000	1000	1000	1275
			16	1000	1000	1750	1800	2325	3500
			18	1000	1200	2100	2175	2725	DR
	2	10	9	1000	1000	1025	1075	1550	2500
			16	1525	2025	3125	3200	3900	DR
			18	1875	2400	3575	3700	DR	DR
	2	12	9	1000	1200	2075	2125	2750	4000
			16	2600	3200	DR	DR	DR	DR
			18	3175	3850	DR	DR	DR	DR
	4	12	9	1775	2350	3500	3550	DR	DR
			16	4175	DR	DR	DR	DR	DR
2x6 Stud Grade	2	12	9	1000	1000	1325	1375	1750	2550
			16	1650	2050	2925	3000	3500	DR
			18	2025	2450	3425	3500	4100	DR
	4	12	9	1125	1500	2225	2275	2775	3800
			16	2650	3150	DR	DR	DR	DR
			18	3125	3675	DR	DR	DR	DR

a. DR = design required
b. Strap shall be installed in accordance with manufacturer's recommendations.

The section of the 2009 IRC covering Method CS-PF (continuous portal frame) is reproduced below:

> **R602.10.4.1.1 Continuous portal frame.** *Continuous portal frame braced wall panels shall be constructed in accordance with Figure R602.10.4.1.1. The number of continuous portal frame panels in a single braced wall line shall not exceed four. For purposes of resisting wind pressures acting perpendicular to the wall, the requirements of Figure R602.10.4.1.1 and IRC Table R602.10.4.1.1 shall be met. There shall be a maximum of two braced wall segments per header and header length shall not exceed 22 feet (6706 mm). Tension straps shall be installed in accordance with the manufacturer's recommendations.*

Note that the pony wall information in IRC Table R602.10.4.1.1 may be used for pony walls over any header within the scope of the table. For conventionally framed headers, the table is slightly conservative, but it is the only prescriptive guidance available for pony walls over headers. The use of this table will ensure that the header/pony wall assembly has sufficient stiffness to prevent a hinge from forming at the header-to-pony wall joint when subjected to wind loads acting against the wall.

Description of continuous structural fiberboard sheathing bracing method

New to the 2009 IRC is bracing Method CS-SFB (continuous structural fiberboard sheathing). It is very similar to Method CS-WSP (continuous wood structural panel sheathing) except that there are restrictions as to where it may be used. Note that Method CS-G (wood structural panel adjacent to garage door opening and supporting roof loads only) and Method CS-PF (continuous portal frame) are based on the use of wood structural panel sheathing and are not applicable for Method CS-SFB.

Section R602.10.5 of the 2009 IRC requires that, when using Method CS-SFB, the entire length of the braced wall line (including areas above and below openings and gable ends, if applicable) must be fully sheathed with a minimum of 1/2-inch structural fiberboard sheathing.

Method CS-SFB (continuous structural fiberboard sheathing)

TABLE 5.10 was developed for this publication in anticipation that future editions of the IRC will include this information in a similar format. It is an interpretation of the code, intended to improve the reader's understanding.

TABLE 5.10

Method CS-SFB (continuous structural fiberboard sheathing)

Method	Material	Minimum Thickness	Figure	Connection Criteria
CS-SFB	Continuous structural fiberboard sheathing	1/2" or 25/32" for 16" stud spacing only		1-1/2" galvanized roofing nails or 8d common (2-1/2" x 0.131") nails at 3" spacing (panel edges) at 6" spacing (intermediate supports)

As with Method CS-WSP, other bracing methods may not be used in a Method CS-SFB wall line. Mixing bracing methods is described in Section R602.10.1.1 of the 2009 IRC and will be addressed later in this chapter.

Also, like Method CS-WSP wall lines, Method CS-SFB wall lines are permitted on a line by line basis in the 2009 IRC and a number of requirements for corner details are provided. Method CS-SFB can also be placed away from corners, as is permitted for intermittent methods. This is permitted in the exception to IRC Section R602.10.5.3 but is contingent on the use of at least one of a couple of detailing provisions. A number of figures have been added to the 2009 IRC to clarify the corner attachment requirements of the continuous sheathing provisions (see **FIGURES 5.16** through **5.19** for illustrations of corner details). Additional information regarding corner details is provided in **CHAPTER 7**.

IRC Table R602.10.5.2 (duplicated in **TABLE 5.11**) was added to the 2009 IRC to provide minimum length requirements for fiberboard sheathing used in Method CS-SFB bracing for various wall heights. Note that Method CS-SFB may only be used in walls up to 10 feet high, and that the minimum panel length is aspect-ratio linked (the minimum bracing length increases as the wall height increases). This is done to ensure that as the wall gets taller, it remains sufficiently stiff.

There are some limitations on the use of Method CS-SFB bracing that set it apart from Method CS-WSP. CS-SFB is not permitted to be used in areas with wind speeds of greater than 100 mph or in SDC D_0, D_1 or D_2. Method CS-SFB bracing is not permitted for use in walls over 10 feet high. Return corners and narrow wall segments adjacent to a corner without a hold down shall be not less than 32 inches (as shown in **FIGURES 5.16** through **5.19**).

TABLE 5.11

Mininum length requirements for Method CS-SFB (continuous structural fiberboard sheathing)[a]

IRC Table R602.10.5.2

Minimum length of structural fiberboard braced wall panel (inches)			Minimum opening clear height next to the structural fiberboard braced wall panel (% of wall height)
8-foot wall	9-foot wall	10-foot wall	
48	54	60	100
32	36	40	85
24	27	30	67

For SI: 1 inch = 25.4 mm, 1 foot = 305 mm
a. Interpolation shall be permitted.

The sections of the 2009 IRC covering Method CS-SFB are reproduced below:

R602.10.5 Continuously sheathed braced wall line using Method SC-SFB (structural fiberboard sheathing). *Continuously sheathed braced wall lines using structural fiberboard sheathing shall comply with this section. Different bracing methods shall not be permitted within a continuously sheathed braced wall line. Other bracing methods prescribed by this code shall be permitted on other braced wall lines on the same story level or on different story levels of the building.*

R602.10.5.1 Continuously sheathed braced wall line requirements. *Continuously-sheathed braced wall lines shall be in accordance with Figure R602.10.4.2 and shall comply with all of the following requirements:*

1. *Structural fiberboard sheathing shall be applied to all exterior sheathable surfaces of a braced wall line including areas above and below openings.*

2. *Only full-height or blocked braced wall panels shall be used for calculating the braced wall length in accordance with Tables R602.10.1.2(1) and R602.10.1.2(2).*
(See **TABLES 6.1** and **6.2**)

R602.10.5.2 Braced wall panel length. *In a continuously sheathed structural fiberboard braced wall line, the minimum braced wall panel length shall be in accordance with Table R602.10.5.2.* (Shown in **TABLE 5.6**.)

R602.10.5.3 Braced wall panel location and corner construction. *A braced wall panel shall be located at each end of a continuously sheathed braced wall line. A minimum 32-inch (813 mm) structural fiberboard sheathing panel corner return shall be provided at both ends of a continuously sheathed braced wall line in accordance with Figure R602.10.4.4(1) In lieu of the corner return, a hold-down device with a minimum uplift design value of 800 lb (3560 N) shall be fastened to the corner stud and to the foundation or framing below in accordance with Figure R602.10.4.4(3).*

Exception: *The first braced wall panel shall be permitted to begin 12-feet 6-inches (3810 mm) from each end of the braced wall line in Seismic Design Categories A, B, and C provided one of the following is satisfied:*

1. *A minimum 32-inch-long (813 mm), full-height structural fiberboard sheathing panel is provided at both sides of a corner constructed in accordance with Figure R602.10.4.4(1) at the braced wall line ends in accordance with Figure R602.10.4.4(4), or*

2. *The braced wall panel closest to the corner shall have a hold-down device with a minimum uplift design value of 800 lb (3560 N) fastened to the stud at the edge of the braced wall panel closest to the corner and to the foundation or framing below in accordance with Figure R602.10.4.4(5).*

R602.10.5.4 Continuously sheathed braced wall lines. *Where a continuously sheathed braced wall line is used in Seismic Design Categories D_0, D_1, and D_2 or regions where the basic wind speed exceeds 100 miles per hour (45 m/s), the braced wall line shall be designed in accordance with accepted engineering practice and the provisions of the International Building Code. Also all other exterior braced wall lines at exterior walls in the same story shall be continuously sheathed.*

Prefabricated bracing panels and ICC-ES reports

FIGURE 5.23

Sample ICC-ES report

In addition to the methods provided in the IRC, there are additional bracing options available that are recognized as code-compliant through reports issued by ICC Evaluation Service (ICC-ES) or other ISO-65 certified agencies that recognize new and innovative building methods and materials as meeting the intended requirements of the building codes. ICC-ES and other similar agencies provide the building official with the assurance that products that fall outside of the building code "umbrella" meet the performance and quality requirements of other code-conforming products. Braces and bracing methods approved for use based on such reports must be installed in strict accordance with all details and limitations included in the report and manufacturer's recomendations.

Code reports

Copies of code-evaluation reports are available at www.icc-es.org or other agencies' websites.

ES REPORT™

ESR-4802

Issued March 1, 2008

This report is subject to re-examination in one year.

ICC Evaluation Service, Inc.
www.icc-es.org

Business/Regional Office ■ 5360 Workman Mill Road, Whittier, California 90601 ■ (562) 699-0543
Regional Office ■ 900 Montclair Road, Suite A, Birmingham, Alabama 35213 ■ (205) 599-9800
Regional Office ■ 4051 West Flossmoor Road, Country Club Hills, Illinois 60478 ■ (708) 799-2305

DIVISION: 07—THERMAL AND MOISTURE PROTECTION
Section: 07410—Metal Roof and Wall Panels

REPORT HOLDER:

ACME CUSTOM-BILT PANELS
52380 FLOWER STREET
CHICO, MONTANA 43820
(808) 664-1512
www.custombiltpanels.com

EVALUATION SUBJECT:

CUSTOM-BILT STANDING SEAM METAL ROOF PANELS:
CB-150

1.0 EVALUATION SCOPE

Compliance with the following codes:

■ 2006 *International Building Code*® (IBC)

■ 2006 *International Residential Code*® (IRC)

Properties evaluated:

■ Weather resistance

■ Fire classification

■ Wind uplift resistance

2.0 USES

Custom-Bilt Standing Seam Metal Roof Panels are steel panels complying with IBC Section 1507.4 and IRC Section R905.10. The panels are recognized for use as Class A roof coverings when installed in accordance with this report.

3.0 DESCRIPTION

3.1 Roofing Panels:

Custom-Bilt standing seam roof panels are fabricated in steel and are available in the CB-150 and SL-1750 profiles. The panels are roll-formed at the jobsite to provide the standing seams between panels. See Figures 1 and 3 for panel profiles.

The standing seam roof panels are roll-formed from minimum No. 24 gage [0.024 inch thick (0.61 mm)] cold-formed sheet steel. The steel conforms to ASTM A 792, with an aluminum-zinc alloy coating designation of AZ50.

3.2 Decking:

Solid or closely fitted decking must be minimum $^{15}/_{32}$-inch-thick (11.9 mm) wood structural panel or lumber sheathing, complying with IBC Section 2304.7.2 or IRC Section R803, as applicable.

4.0 INSTALLATION

4.1 General:

Installation of the Custom-Bilt Standing Seam Roof Panels must be in accordance with this report, Section 1507.4 of the IBC or Section R905.10 of the IRC, and the manufacturer's

published installation instructions. The manufacturer's installation instructions must be available at the jobsite at all times during installation.

The roof panels must be installed on solid or closely fitted decking, as specified in Section 3.2. Accessories such as gutters, drip angles, fascias, ridge caps, window or gable trim, valley and hip flashings, etc., are fabricated to suit each job condition. Details must be submitted to the code official for each installation.

4.2 Roof Panel Installation:

4.2.1 CB-150: The CB-150 roof panels are installed on roofs having a minimum slope of 2:12 (17 percent). The roof panels are installed over the optional underlayment and secured to the sheathing with the panel clip. The clips are located at each panel rib side lap spaced 6 inches (152 mm) from all ends and at a maximum of 4 feet (1.22 m) on center along the length of the rib, and fastened with a minimum of two No. 10 by 1-inch pan head corrosion-resistant screws. The panel ribs are mechanically seamed twice, each pass at 90 degrees, resulting in a double-locking fold.

4.3 Fire Classification:

The steel panels are considered Class A roof coverings in accordance with the exception to IBC Section 1505.2 and IRC Section R902.1.

4.4 Wind Uplift Resistance:

The systems described in Section 3.0 and installed in accordance with Sections 4.1 and 4.2 have an allowable wind uplift resistance of 45 pounds per square foot (2.15 kPa).

5.0 CONDITIONS OF USE

The standing seam metal roof panels described in this report comply with, or are suitable alternatives to what is specified in, those codes listed in Section 1.0 of this report, subject to the following conditions:

5.1 Installation must comply with this report, the applicable code, and the manufacturer's published installation instructions. If there is a conflict between this report and the manufacturer's published installation instructions, this report governs.

5.2 The required design wind loads must be determined for each project. Wind uplift pressure on any roof area must not exceed 45 pounds per square foot (2.15 kPa).

6.0 EVIDENCE SUBMITTED

Data in accordance with the ICC-ES Acceptance Criteria for Metal Roof Coverings (AC166), dated October 2007.

7.0 IDENTIFICATION

Each standing seam metal roof panel is identified with a label bearing the product name, the material type and gage, the Acme Custom-Bilt Panels name and address, and the evaluation report number (ESR-4802).

ANSI

ANSI Accredited Program
PRODUCT CERTIFICATION

Mixing bracing methods

While the 2006 IRC was silent on the subject of mixing bracing methods, the 2009 IRC addresses the issue in Section R602.10.1.1.

> **R602.10.1.1 Braced wall panels.** *Braced wall panels shall be constructed in accordance with the intermittent bracing methods specified in Sections R602.10.2, or the continuous sheathing methods specified in Section R602.10.4 and R602.10.5. Mixing of bracing method shall be permitted as follows:*
>
> 1. *Mixing bracing methods from story to story is permitted.*
>
> 2. *Mixing bracing methods from braced wall line to braced wall line within a story is permitted, except that continuous sheathing methods shall conform to the additional requirements of Section R602.10.4 and R602.10.5.*
>
> 3. *Mixing bracing methods within a braced wall line is permitted only in Seismic Design Categories A and B, and detached dwellings in Seismic Design Category C. The length of required bracing for the braced wall line with mixed sheathing types shall have the higher bracing length requirement, in accordance with IRC Tables R602.10.1.2(1) and R602.10.1.2(2), of all types of bracing used.*

In other words:

1. Story-to-story

Mixing bracing methods from story-to-story is permitted. An example is using Method WSP (wood structural panel) bracing on the bottom of a three-story building while using Method PBS (particleboard sheathing) or Method GB (gypsum board) bracing on the top and/or second story.

2. Braced wall line to braced wall line on a given story

A builder is permitted to use different bracing methods on different walls within a story. For example, in a "window wall", a designer may use Method WSP (wood structural panel) bracing along one wall line of a story because its lower bracing length requirements can help to accommodate additional windows. On the other walls having fewer windows, the designer may use method Method GB (gypsum board) or other bracing method.

Mixing bracing methods with Method CS-WSP (continuous wood structural panel sheathing) must further comply with the additional requirements of the exception to IRC Section R602.10.4:

> **Exception:** *Within Seismic Design Categories A, B and C or in regions where the basic wind speed is less than or equal to 100 miles per hour (45 m/s), other bracing methods prescribed by this code shall be permitted on other braced wall lines on the same story level or on any braced wall line on different story levels of the building.*

This exception specifically permits, in areas of low or moderate seismicity or in regions where the wind speeds are 100 mph or less, continuous sheathing bracing methods to be used with other bracing methods on other walls of the same story and on other levels. What is inferred by this section is that in high seismic or high wind zones, mixing intermittent bracing methods with Method CS-WSP (continuous wood structural panel sheathing) is not permitted on the same story or other levels. Note that Method CS-SFB is not permitted in high seismic or high wind zones.

For the 2009 IRC, it was determined that in high seismic and high wind zones, all exterior walls of a given level must be continuously sheathed if a single wall on that level is continuously sheathed. However, walls inside the building or walls on other levels may use whatever bracing meets the requirements for that wall or level. This was more clearly stated in the 2007 Supplement to the IRC:

> **R602.10.4.7 (2007 Supplement) Continuously sheathed braced wall lines.** *Where a continuously sheathed braced wall line is used in Seismic Design Categories D_0, D_1 and D_2 or regions where the basic wind speed exceeds 100 miles per hour, all other exterior braced wall lines in the same story shall be continuously sheathed.*

Provisions covering mixing of bracing methods when using Method CS-SFB (continuous structural fiberboard sheathing) are covered briefly in IRC Section R602.10.5:

> **R602.10.5 Continuously sheathed braced wall line using Method SC-SFB (structural fiberboard sheathing).** *...Other bracing methods prescribed by this code shall be permitted on other braced wall lines on the same story level or on different story levels of the building...*

3. Mixing in one braced wall line

For one- and two-family dwellings in SDC A, B and C, and for townhouses in SDC A and B, mixing bracing methods within a braced wall line is permitted. The length of the required bracing for the braced wall line with mixed sheathing types must have the greatest required bracing length, per IRC Tables R602.10.1.2(1) and (2), of all types of bracing used in that wall line.

Wall lines using Method CS-WSP (continuous wood structural panel sheathing) or Method CS-SFB (continuous structural fiberboard sheathing) bracing shall conform to the following:

a. Different bracing methods shall not be permitted within a continuously sheathed braced wall line.

b. Other bracing methods shall be permitted on other braced wall lines on the same story or on different stories of the building.

c. Where a continuously sheathed braced wall line (CS-WSP, CS-G and CS-PF) is used in SDC D_0, D_1 or D_2, or regions where the basic wind speed exceeds 100 mph, all other exterior braced wall lines on the same story shall be continuously sheathed with wood structural panels. Note that Method CS-SFB is not permitted in SDC D_0, D_1 or D_2.

Checklist: Using bracing provisions of Chapter 5

A. Panel width – Is each bracing panel the right minimum length? See **TABLE 5.12** for minimum length guidelines.

TABLE 5.12

Length requirements for braced wall panels

Method		Minimum Length					Contributing Length
		Wall Height					
		8'	9'	10'	11'	12'	
DWB , WSP, SFB, PBS, PCP, HPS, GB (double-sided)		48"	48"	48"	53"	58"	Actual[1]
GB (single-sided)		96"	96"	96"	106"	116"	Actual[1]
ABW	SDC A, B and C, wind speed < 110 mph	28"	32"	34"	38"	42"	48"
	SDC D_0, D_1 and D_2, wind speed < 110 mph	32"	32"	34"	NP	NP	48"
PFH	One-story	16"	16"	16"	18"	20"	48"
	Two-story	24"	24"	24"	27"	29"	48"
PFG	Garage	24"	24"	24"	NP	NP	1.5 x Actual[1]
CS-WSP, CS-SFB	Adjacent clear opening height						
	≤ 64"	24"	27"	30"	33"[2]	36"[2]	
	68"	26"	27"	30"	33"[2]	36"[2]	
	72"	27"	27"	30"	33"[2]	36"[2]	
	76"	30"	29"	30"	33"[2]	36"[2]	
	80"	32"	30"	30"	33"[2]	36"[2]	
	84"	35"	32"	32"	33"[2]	36"[2]	
	88"	38"	35"	33"	33"[2]	36"	
	92"	43"	37"	35"	35"[2]	36"[2]	
	96"	48"	41"	38"	36"[2]	36"[2]	
	100"		44"	40"	38"[2]	38"[2]	Actual[1]
	104"		49"	43"	40"[2]	39"[2]	
	108"		54"	46"	43"[2]	41"[2]	
	112"			50"	45"[2]	43"[2]	
	116"			55"	48"[2]	45"[2]	
	120"			60"	52"[2]	48"[2]	
	124"				56"[2]	51"[2]	
	128"				61"[2]	54"[2]	
	132"				66"[2]	58"[2]	
	136"					62"[2]	
	140"					66"[2]	
	144"					72"[2]	
CS-G		24"	27"	30"	33"	36"	Actual[1]
CS-PF		16"	18"	20"	22"	24"	Actual[1]

For SI: 1 in. = 25.4 mm
NP = Not permitted
1. Actual length when greater than or equal to the minimum length.
2. Heights above 10 ft are not permitted for Method CS-SFB.

B. Have intermittent braced panels been used that are less than 48 inches in length? If so, have the effective lengths been adjusted in accordance with IRC Table R602.10.3?

C. Is the stud height less than or equal to 12 feet? (See *TABLE 5.12* for methods that are limited to 10 feet.)

D. Is the story height less than or equal to 139 inches for a stud height of 10 feet or less and 163 inches for a stud height of 12 feet?

E. Bracing material attachment – are bracing materials attached with the connection criteria in accordance with IRC Table R602.10.2?

F. Are gypsum panels, when used for bracing, attached with nails or screws at 7 inches on center at all edges, including top and bottom plates?

G. Is gypsum board applied on the backside of all braced wall panels? It shall be nailed in accordance with IRC Table 702.3.5 for interior gypsum wall board. If gypsum wall board is not applied, have the bracing lengths been increased in accordance with Footnote f of IRC Table R602.10.1.2(1) (1.4 increase) or IRC Section R602.10.2.1, Exception 3, (1.5 increase) for wind bracing? For seismic bracing, the adjustment of 1.5 as specified in IRC Section R602.10.2.1, Exception 3, is applicable. (Note that the ICC Staff is aware that there is a conflict between the 1.5 and 1.4 adjustment requirements for wind bracing. As it stands, either one can be used, as both are specified by the code. This will hopefully be corrected in the 2012 edition of the IRC.)

H. Panel grade and thickness – are bracing panels the proper grade and thickness, in accordance with IRC Table R602.10.2?

I. Have the bracing lengths for wind and seismic bracing been adjusted for length in accordance with the footnotes to IRC Table R602.10.1.2(1) for wind and IRC Table R602.10.1.2(3) for seismic bracing?

J. Is the wall bracing present in all wall lines the greater of that required in IRC Table R602.10.1.2(1) for wind bracing and R602.10.1.2(2) for seismic bracing? Is the length of bracing in any braced wall line 48 inches or greater (IRC Section R602.10.1.2)?

K. Are bracing Methods ABW or PFH used (IRC Table R602.10.2)?
  - If so, are hold downs sized and installed properly?

L. Are continuous sheathing provisions (Methods CS-WSP, CS-G, CS-PF, CS-SFB) used (IRC Section R602.10.4 and R602.10.5)? Is the continuously sheathed wall braced with wood structural panels or structural fiberboard sheathing above and below openings, including gable ends?
  - If so, are panel lengths in accordance with *TABLE 5.12* length requirements for braced wall panels of this publication?
  - Is nailing used in corners in accordance with IRC Figure R602.10.4.4(1)?
  - Is the portal frame (Method CS-PF), if used, built in accordance with IRC Figure R602.10.4.1.1, and does it occur in a continuous wood structural panel sheathing wall line?

M. Are braced wall lines spaced properly?
  - Less than or equal to 60 feet for one- and two-family dwellings in SDC A, B and C and townhouses in SDC A and B.

- Less than or equal to 35 feet (increase to 50 feet per IRC Table R602.10.1.2(3)) for townhouses in SDC C.

- Less than or equal to 25 feet (increase to 35 feet per IRC Table R602.10.1.5 if used in conjunction with a single 900 sf room) for one- and two-family dwellings SDC D_0, D_1 and D_2.

N. Are braced wall lines connected to foundation in accordance with IRC Section R403.1.6, except as required below:

- SDC D_0, D_1 and D_2 and townhouses in SDC C: Are the provisions of IRC Section R602.11.1 met?

- Do stepped footings in SDC D_0, D_1 and D_2 meet the additional requirements of IRC Section R602.11.2?

O. Are masonry foundation walls supporting bracing panels 48 inches or less in height and length reinforced in accordance with IRC Figure R602.10.7?

P. Are top plate lap splices detailed in accordance with IRC Table R602.3(1), Item 13; or Section R602.10.6.1 (eight 16d, each side of splice)?

Q. Are all braced wall panels properly attached top and bottom?

- Connection to floors above and below braced wall panels: attached in accordance with IRC Section R602.10.6.

- Connection to roof framing: attached in accordance with IRC Section R602.10.6.2.

R. Are braced wall panels spaced away from the end of the wall line?

- For the placement of intermittent braced wall panels in SDC A, B and C, the combined distance to the first braced wall panel at both ends is 12 feet 6 inches. For the continuous sheathing methods, braced wall panels can be located 12 feet 6 inches from each end of the braced wall line, which is a combined distance of 25 feet.

- For SDC D_0, D_1 and D_2, braced walls are required to have a braced wall panel at each end. An exception to having a braced wall panel at the end of the wall line is given for Method WSP and CS-WSP per IRC Section R602.10.1.4.1. This exception allows a braced wall panel to be located 8 feet from each end of the braced wall line, provided one of the following is satisfied:

  1. Minimum 24-inch bracing panel applied to both sides of corner at intersection walls and attached in accordance with IRC Figure R602.10.1.4.4.

  2. A minimum 1800-lb hold down used at the end of each braced wall panel closest to the corner.

S. Are bracing panels offset from braced wall line?

- A maximum total out-to-out offset of 8 feet between wall sections making up a braced wall line is permitted. A maximum 4-foot offset to each side of the designated wall line is permitted. The designated braced wall line does not have to fall on a real wall line.

T. Do connections at exterior braced walls that support roof trusses or rafters meet the uplift load path requirements of IRC Section R602.10.1.2.1?

U. For dwellings in SDC D_0, D_1 and D_2 and townhouses in SDC C , are there irregularities per IRC Section R301.2.2.2.5 present?

CHAPTER 6

How Much Bracing is Needed?

The amount of wall bracing required in each braced wall line, as specified by the building code, depends on the Seismic Design Category (SDC), wind speed, number of stories above the braced wall line and the method of bracing used.

The procedure for calculating bracing has changed in the 2009 IRC. The user must now consult two separate bracing tables – one for wind loads and one for seismic loads – to compute the required amount of bracing. However, if the structure is located in a low seismic area, it may only be necessary to consult the wind bracing table: IRC Section R301.2.2 exempts detached one- and two-family dwellings in SDC A, B and C and townhouses in SDC A and B from the seismic provisions of the IRC.

Another change is that the bracing tables now provide the required bracing length in total feet of bracing, rather than as a percentage of wall line length. This eliminates the need for the user to compute the value.

Summary

Changes in the 2009 IRC

The concept of computing bracing from two separate tables is introduced in IRC Section R602.10.1.2. Also, the following provisions are introduced in IRC Section R602.10.1:

- The requirement for a minimum of 48 inches of bracing for any braced wall line (after all adjustments) (R602.10.1.2).

- Provisions permitting angled walls to be used for bracing (R602.10.1.3). See **CHAPTER 7**.

IRC Section R602.10.1.2 is reproduced below:

> **R602.10.1.2 Length of bracing.** *The length of bracing along each braced wall line shall be the greater of that required by the design wind speed and braced wall line spacing in accordance with Table R602.10.1.2(1) as adjusted by the factors in the footnotes or the Seismic Design Category and braced wall line length in accordance with Table R602.10.1.2(2) as adjusted by the factors in Table R602.10.1.2(3) or braced wall panel location requirements of Section R602.10.1.4. Only walls that are parallel to the braced wall line shall be counted towards the bracing requirement of that line, except angled walls shall be counted in accordance with Section R602.10.1.3. In no case shall the minimum total length of bracing in a braced wall line, after all adjustments have been taken be less than 48 inches total.*

Why two tables? Why have the amounts of bracing changed?

The answer to the first question is as follows. Previous IRC bracing tables assumed a braced wall line spacing of 35 feet for wind speeds of 110 mph or less and SDC A, B and C, and a braced wall line spacing of 25 feet for SDC D_0, D_1 and D_2. Basing the bracing requirement on the width of the building (the braced wall line spacing) is appropriate for wind. As discussed in **CHAPTER 1**, wind pushes against a building in the same manner that it pushes against a sail on a boat. When wind acts on the building width, the length of the building (dimension parallel to the wind) is irrelevant to determining the bracing required to resist that wind load. In other words, a short building that is 35-feet wide has the same "sail area" – and receives the same wind load – as a long building that is 35-feet wide. Therefore, if the short building requires 12 feet of bracing, the long building also requires 12 feet of bracing. This reasoning is contrary to previous IRC bracing tables that based the amount of wind bracing on a percentage of the braced wall line length, which required more bracing for the long building than the short building. These different approaches are illustrated in **FIGURE 6.1**. In the 2009 wind bracing table, IRC Table R602.10.1.2(1) (reproduced in **TABLE 6.1**), the user inputs *braced wall line spacing* to determine the required bracing length in feet, regardless of building length.

FIGURE 6.1

Wall bracing – wind loads

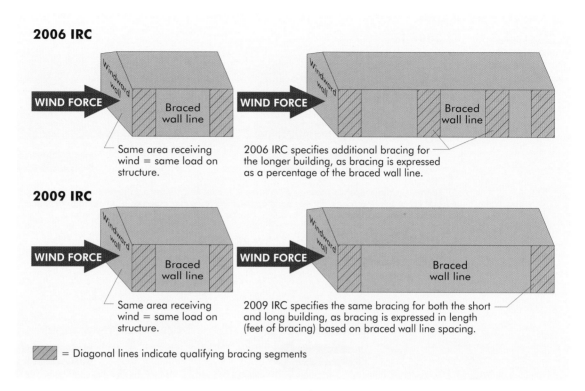

2006 IRC

WIND FORCE · Braced wall line

Same area receiving wind = same load on structure.

2006 IRC specifies additional bracing for the longer building, as bracing is expressed as a percentage of the braced wall line.

2009 IRC

WIND FORCE · Braced wall line

Same area receiving wind = same load on structure.

2009 IRC specifies the same bracing for both the short and long building, as bracing is expressed in length (feet of bracing) based on braced wall line spacing.

= Diagonal lines indicate qualifying bracing segments

There are now two separate tables for computing bracing, because bracing for seismic loads is just the opposite of bracing for wind loads. When determining the amount of bracing required to resist seismic forces, the length of the building parallel to the direction of loading is the most important consideration. This is because mass is generally evenly distributed along the length and width of a building. For a given building width, the long building has more mass – and thus receives greater earthquake forces – than the short building. As a result, the long building requires a greater amount of bracing. For this reason, in the 2009 seismic bracing table (IRC Table R602.10.1.2(2), reproduced in *TABLE 6.2*), the user inputs the *length of the braced wall line* to determine the amount of bracing required. This is illustrated in *FIGURE 6.2*.

Previous joint wind and seismic bracing tables in the IRC were based on seismic loads, so – as with the 2009 IRC seismic bracing table – the amount of required bracing increased as the braced wall line length increased. However, unlike previous editions of the IRC, the amount of required bracing is now provided in feet rather than a percentage of braced wall line length, eliminating the need for the user to compute the necessary feet of bracing.

Since wind and seismic loads act on a structure differently, a single table cannot accommodate both forces of nature (as had been attempted in previous versions of the IRC). The logical solution was to separate wind and seismic loads into two separate bracing tables. In the 2009 IRC, the user must determine bracing length from both tables: the required length of bracing is the greater of the two lengths.

FIGURE 6.2

**Wall bracing –
seismic loads**

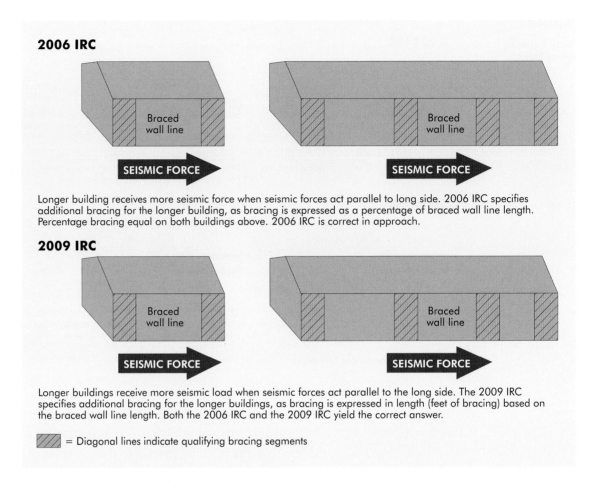

2006 IRC

Braced wall line

SEISMIC FORCE

Braced wall line

SEISMIC FORCE

Longer building receives more seismic force when seismic forces act parallel to long side. 2006 IRC specifies additional bracing for the longer building, as bracing is expressed as a percentage of braced wall line length. Percentage bracing equal on both buildings above. 2006 IRC is correct in approach.

2009 IRC

Braced wall line

SEISMIC FORCE

Braced wall line

SEISMIC FORCE

Longer buildings receive more seismic load when seismic forces act parallel to the long side. The 2009 IRC specifies additional bracing for the longer buildings, as bracing is expressed in length (feet of bracing) based on the braced wall line length. Both the 2006 IRC and the 2009 IRC yield the correct answer.

= Diagonal lines indicate qualifying bracing segments

What is the basis for the new tables?

Although efforts were made to quantitatively fit wind bracing considerations for various SDCs into the IRC's previous joint wind and seismic bracing tables, these tables were technically based only on seismic load considerations. This is why previous bracing tables specified more bracing for both wind and seismic loads as the building grew longer. As discussed previously, determining the bracing amount based on braced wall line length is appropriate for seismic loads, but not for wind loads.

The ICC Ad Hoc Committee on Wall Bracing realized the necessity of developing a rational basis for wind load bracing tables. The committee determined the actual loads that act on a structure within the range of wind speeds and building sizes covered by the IRC. They investigated structural resistance, evaluated and determined the capacities of existing bracing and proposed bracing methods. The committee also considered other factors that would impact the lateral performance of the structure beyond the designated bracing panels, such as the strength contribution of interior walls and finishes, actual building performance compared with calculated performance, construction quality and factors of safety.

The wind bracing table that resulted from the ICC Ad Hoc Committee's efforts was formatted to provide the user with the required bracing amount in total feet of bracing. And although the basis for determining seismic load bracing had not changed, the committee reformatted the seismic bracing table to also yield required bracing in feet, so as to be consistent with the wind bracing table.

TABLES 6.1 and **6.2** are the results of this multi-year project undertaken by a team of building officials, academics, design professionals, builders, structural engineers, ICC staff, product manufacturers and others. Each table includes adjustment factors. These adjustment factors – modifications to the amount of bracing based on variations in the structural geometry – are provided in footnotes and are different for each table. It is important to note that neglecting an adjustment factor can result in insufficient bracing for a specific application. For example, the wind bracing tables are based on a roof eave-to-ridge height of 10 feet. If the roof height of a given single-story structure is 15 feet and Footnote c of the wind bracing table is ignored, the wall bracing will be insufficient by 30 percent. For this reason, all table footnotes must be considered carefully.

How much bracing is needed for wind?

The 2009 IRC wind bracing table (IRC Table R602.10.1.2(1)) and its footnotes are reproduced in **TABLE 6.1**. (Note that in the table, the "Story Location" is represented by green shading in the building icon.)

As illustrated in **FIGURE 6.3**, the information in the wind bracing table is based on:

- Exposure Category B (explained in **CHAPTER 4**)

- 30-foot mean roof height

- 10-foot eave-to-ridge-height

- 10-foot wall height per story

- Two braced wall lines per direction of wind

FIGURE 6.3

Basis for wind bracing table

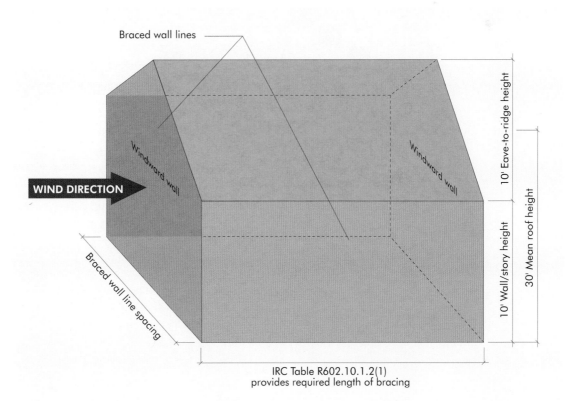

Braced wall lines

Windward wall

WIND DIRECTION

Windward wall

10' Eave-to-ridge height

30' Mean roof height

10' Wall/story height

Braced wall line spacing

IRC Table R602.10.1.2(1)
provides required length of bracing

TABLE 6.1

Bracing requirements based on wind speed (as a function of braced wall line spacing)

IRC Table R602.10.1.2(1)[a,b,c,d,e]

Exposure Category B
30 ft mean roof height
10 ft eave to ridge height
10 ft wall height
2 braced wall lines

Minimum total length (feet) of braced wall panels required along each braced wall line

Basic Wind Speed	Story Location	Braced Wall Line Spacing (ft)	Method LIB[f,h]	Method GB (Double-Sided)[g]	Methods DWB, WSP, SFB, PBS, PCP, HPS[f,i]	Cont. Sheathing
≤ 85 mph		10	3.5	3.5	2.0	1.5
		20	6.0	6.0	3.5	3.0
		30	8.5	8.5	5.0	4.5
		40	11.5	11.5	6.5	5.5
		50	14.0	14.0	8.0	7.0
		60	16.5	16.5	9.5	8.0
		10	6.5	6.5	3.5	3.0
		20	11.5	11.5	6.5	5.5
		30	16.5	16.5	9.5	8.0
		40	21.5	21.5	12.5	10.5
		50	26.5	26.5	15.0	13.0
		60	31.5	31.5	18.0	15.5
		10	NP	9.0	5.5	4.5
		20	NP	17.0	10.0	8.5
		30	NP	24.5	14.0	12.0
		40	NP	32.0	18.0	15.5
		50	NP	39.0	22.5	19.0
		60	NP	46.5	26.5	22.5
≤ 90 mph		10	3.5	3.5	2.0	2.0
		20	7.0	7.0	4.0	3.5
		30	9.5	9.5	5.5	5.0
		40	12.5	12.5	7.5	6.0
		50	15.5	15.5	9.0	7.5
		60	18.5	18.5	10.5	9.0
		10	7.0	7.0	4.0	3.5
		20	13.0	13.0	7.5	6.5
		30	18.5	18.5	10.5	9.0
		40	24.0	24.0	14.0	12.0
		50	29.5	29.5	17.0	14.5
		60	35.0	35.0	20.0	17.0
		10	NP	10.5	6.0	5.0
		20	NP	19.0	11.0	9.5
		30	NP	27.5	15.5	13.5
		40	NP	35.5	20.5	17.5
		50	NP	44.0	25.0	21.5
		60	NP	52.0	30.0	25.5

For SI: 1 foot = 304.8 mm

TABLE 6.1 (CONT.)

Basic Wind Speed	Story Location	Braced Wall Line Spacing (ft)	Method LIB[f,h]	Method GB (Double-Sided)[g]	Methods DWB, WSP, SFB, PBS, PCP, HPS[f,i]	Cont. Sheathing
	Exposure Category B 30 ft mean roof height 10 ft eave to ridge height 10 ft wall height 2 braced wall lines			**Minimum total length (feet) of braced wall panels required along each braced wall line**		
≤ 100 mph		10	4.5	4.5	2.5	2.5
		20	8.5	8.5	5.0	4.0
		30	12.0	12.0	7.0	6.0
		40	15.5	15.5	9.0	7.5
		50	19.0	19.0	11.0	9.5
		60	22.5	22.5	13.0	11.0
		10	8.5	8.5	5.0	4.5
		20	16.0	16.0	9.0	8.0
		30	23.0	23.0	13.0	11.0
		40	29.5	29.5	17.0	14.5
		50	36.5	36.5	21.0	18.0
		60	43.5	43.5	25.0	21.0
		10	NP	12.5	7.5	6.0
		20	NP	23.5	13.5	11.5
		30	NP	34.0	19.5	16.5
		40	NP	44.0	25.0	21.5
		50	NP	54.0	31.0	26.5
		60	NP	64.0	36.5	31.0
< 110 mph		10	5.5	5.5	3.0	3.0
		20	10.0	10.0	6.0	5.0
		30	14.5	14.5	8.5	7.0
		40	18.5	18.5	11.0	9.0
		50	23.0	23.0	13.0	11.5
		60	27.5	27.5	15.5	13.5
		10	10.5	10.5	6.0	5.0
		20	19.0	19.0	11.0	9.5
		30	27.5	27.5	16.0	13.5
		40	36.0	36.0	20.5	17.5
		50	44.0	44.0	25.5	21.5
		60	52.5	52.5	30.0	25.5
		10	NP	15.5	9.0	7.5
		20	NP	28.5	16.5	14.0
		30	NP	41.0	23.5	20.0
		40	NP	53.0	30.5	26.0
		50	NP	65.5	37.5	32.0
		60	NP	77.5	44.5	37.5

For SI: 1 foot = 304.8 mm

The footnotes to the wind bracing table include adjustment factors to accommodate variations from the above assumptions for the wide range of residential structures covered by the IRC. Where appropriate, these footnotes are explained in the text below. Not all of these adjustments are new to the IRC. Many have been taken from other sections and collected into a common location in the 2009 edition.

> **Footnote a:** *Tabulated bracing lengths are based on Wind Exposure Category B, a 30 ft mean roof height, a 10 ft eave to ridge height, a 10 ft wall height, and two braced wall lines sharing load in a given plan direction on a given story level. Methods of bracing shall be as described in Sections R602.10.2, R602.10.4 and R602.10.5. Interpolation shall be permitted.*

> **Footnote b:** *For other mean roof heights and exposure categories, the required bracing length shall be multiplied by the appropriate factor from the following table:*

Number of Stories	Exposure/Height Factors		
	Exposure B	Exposure C	Exposure D
1	1.0	1.2	1.5
2	1.0	1.3	1.6
3	1.0	1.4	1.7

Note that the wind bracing table is based on Exposure Category B. The Exposure Category for a given jurisdiction can be obtained from the local building department. (Refer to Table R301.2(1), shown in **TABLE 4.3**, of the locally adopted version of the IRC.)

As the number of stories increases (increasing the mean roof height), the adjustment factors increase. This is to accommodate the increased exposure and larger "sail area" of the structure.

> **Footnote c:** *For other roof-to-eave ridge heights, the required bracing length shall be multiplied by the appropriate factor from the following table:*

Support Condition	Roof Eave-to-Ridge Height			
	5 ft	10 ft	15 ft	20 ft
Roof Only	0.7	1.0	1.3	1.6
Roof + Floor	0.85	1.0	1.15	1.3
Roof + 2 Floors	0.9	1.0	1.1	NP

Interpolation shall be permitted.

FIGURE 6.4

**Eave-to-ridge
height**

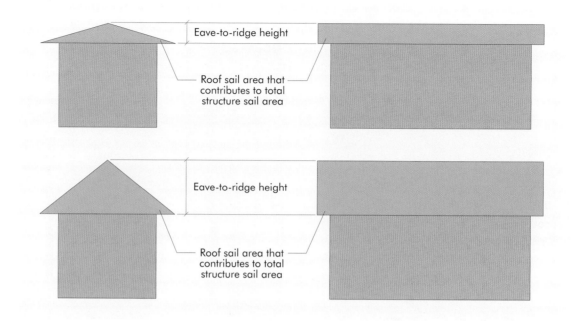

Eave-to-ridge height is an important consideration because it increases the sail area of a structure, therefore increasing the wind load on the structure, as illustrated in **FIGURE 6.4**. Increasing eave-to-ridge height by as little as 5 feet, from 10 to 15 feet for example, can increase the required bracing panel length by up to 30 percent. The first column in the Footnote c table, *Support Condition*, refers to the construction that exists above the wall line for which the amount of bracing is being determined. Note that as the number of stories increases (the support conditions) the contribution of the roof eave-to-ridge height to the sail area decreases.

> **Footnote d:** *For a maximum 9-foot wall height, the table values shall be permitted to be multiplied by 0.95. For a maximum 8-foot wall height, the table values shall be permitted to be multiplied by 0.90. For a maximum 12-foot wall height, the table values shall be multiplied by 1.1.*

Like eave-to-ridge height, increasing the sail area of the structure by making it taller increases the wind load on the structure, thus requiring more bracing. Because the bracing wind table assumes a 10-foot wall height, taller walls require an adjustment to increase the amount of required bracing, while shorter walls are permitted an adjustment to decrease the amount of required bracing. This adjustment can be applied to each story individually. For example, in a house with a first story wall height of 9 feet and a second story wall height of 8 feet, the bracing amount for the first story can be reduced by 5 percent, and the bracing amount for the second story reduced by 10 percent.

Footnote e: *For three or more braced wall lines in a given plan direction, the required bracing length on each braced wall line shall be multiplied by the appropriate factor from the following table (see **FIGURE 6.5**):*

Number of Braced Wall Lines	Adjustment Factor
3	1.30
4	1.45
≥5	1.60

FIGURE 6.5

Braced wall line spacing

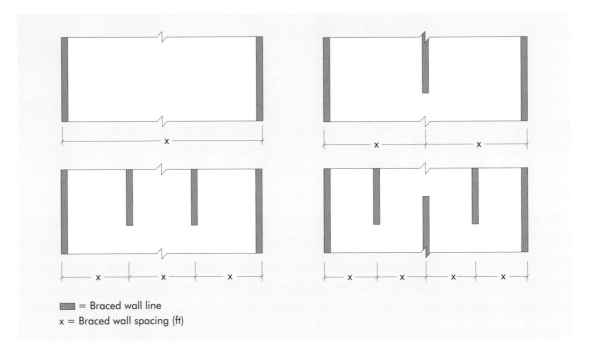

= Braced wall line

x = Braced wall spacing (ft)

A first glance at Footnote e may give the impression that adding additional interior bracing walls is a disadvantage because the amount of bracing in the wall lines must increase by an adjustment factor. However, the values in IRC Table R602.10.1.2(1) are actually based on the spacing of the braced wall lines. If, for example, the distance between two exterior walls in a home design requires a length of bracing that cannot be accommodated by the exterior wall lines alone, a braced wall line may be added through the interior of the structure. In this case, because the spacing of the braced wall lines decreases, the amount of bracing required for the braced wall lines also decreases. The adjustment factor is applied to the required bracing based on this <u>reduced</u> amount of bracing for the braced wall lines. This method is conservative.

Footnote f: *Bracing lengths are based on the application of gypsum board finish (or equivalent) applied to the inside face of a braced wall panel. When gypsum board finish (or equivalent) is not applied to the inside face of braced wall panels, the tabulated lengths shall be multiplied by the appropriate factor from the following table:*

Bracing Method	Adjustment Factor
Method LIB	1.8
Methods DWB, WSP, SFB, PBS, PCP and HPS	1.4

The wind bracing table was developed based on balancing the wind load acting on the structure against the strength of the bracing material providing resistance. The Ad Hoc Committee on Wall Bracing assumed that gypsum wall board would likely be applied on the inside surface of braced wall panels. The addition of gypsum wall board, even though not attached with the same quantity of fasteners as Method GB (gypsum board) bracing, does add strength and stiffness to the bracing; therefore, in cases in which gypsum board is not applied on the inside surface of braced wall panels, an adjustment must be made. In modern residential construction, the absence of gypsum board finish material is only likely to occur at gable end walls (above the plate) and at exterior garage walls. Typically, in either of these applications, it is not difficult to increase the amount of bracing because these walls are not likely to have many openings. Note that Method GB is not listed in the Bracing Method column in the table above. The required length for Method GB (for both single- and double-sided applications) is covered in IRC Section R602.10.3.

Footnote g: *Bracing lengths for Method GB (gypsum board) are based on the application of gypsum board on both faces of a braced wall panel. When Method GB bracing is provided on only one side of the wall, the required bracing amounts shall be doubled. When Method GB braced wall panels installed in accordance with Section R602.10.2 are fastened at 4 inches on center at panel edges, including top and bottom plates, and are blocked at all horizontal joints, multiplying the required bracing percentage for wind loading by 0.7 shall be permitted.*

Footnote g refers to the wind bracing table column labeled *Method GB (Double-Sided)*. The lengths in this column assume that Method GB bracing is applied to both sides of the braced wall line. If Method GB bracing is installed on only one side of the braced wall line, this footnote, requiring the bracing amounts to be doubled, applies. For example, if an exterior braced wall line is braced with Method GB only on the inside of the wall, the length value provided in the *Method GB (Double-Sided)* column of IRC Table R602.10.1.2(1) must be doubled.

Using the lengths as stated from the *Method GB (Double-Sided)* column requires that both sides of the braced wall line are constructed to meet the requirements of Method GB bracing, including type and quantity of fasteners. Fastener spacing must be 7 inches on center at all panel ends, edges and intermediate supports. Note that the typical attachment for gypsum board installed as a wall covering (IRC Section R702.3.5) is not sufficient to provide the strength required for Method GB bracing.

Footnote g permits the required length of Method GB bracing to be multiplied by a factor of 0.7 for wind applications only.

> **Footnote h:** *Method LIB (let-in bracing) shall have gypsum board attached to at least one side according to the Section R602.10.2 Method GB (gypsum board) requirements.*

While all intermittent bracing methods shall have gypsum board on at least one side of the wall (IRC Section R602.10.2.1), Footnote h requires that Method LIB (let-in bracing) have gypsum board installed in accordance with Method GB bracing at the Method LIB location on either side of the wall. The other bracing methods permit the gypsum board to be installed as a wall covering rather than a Method GB bracing material. When gypsum board is installed as Method GB bracing, the fasteners must be spaced 7 inches on center at all panel ends, edges and intermediate supports. The fastener schedule for gypsum board installed as a wall covering (IRC Section R702.3.5) is far less stringent than when it is installed as a bracing material.

> **Footnote i:** *Required bracing length for Methods DWB, WSP, SFB, PBS, PCP, and HPS in braced wall lines located in one-story buildings and in the top story of two- or three-story buildings shall be permitted to be multiplied by 0.80 when an approved tie-down device with a minimum uplift design value of 800 pounds (3560 N) is fastened to the end studs of each braced wall panel in the braced wall line and to the foundation or framing below.*

Footnote i is a new concept for the 2009 IRC. In developing the new wind bracing provisions, braced wall panels only supporting the roof above them (one-story buildings and the top story of multi-story buildings) were recognized to be only partially effective in resisting wind forces. The use of an 800 pound hold-down device, installed in accordance with the manufacturer's recommendations, increases the capacity of bracing methods when supporting roof loads only. A 20 percent reduction in the length of bracing is permitted due to the increased capacity achieved by adding a hold-down device. Note that the hold downs must be used at both ends of every bracing panel in the wall line and the hold-down anchorage must extend through all lower floors (if any) to provide a continuous load path until anchored into the foundation.

How much bracing is needed for seismic?

For the 2009 IRC, the seismic bracing table has undergone a major formatting revision with minor changes to the actual content. The table has been reworked to provide the amount of required bracing in total feet of bracing length instead of a percentage of braced wall line length. This revision was accomplished by basing the table on the length of the braced wall line and fixing the braced wall line spacing at 25 feet.

The 2009 IRC seismic bracing table (IRC Table R602.10.1.2(2)) is reproduced in **TABLE 6.2**. (Note that in the table, the "Story Location" is represented by green shading in the building icon.) IRC Section R301.2.2 exempts detached one- and two-family dwellings in SDC A, B and C and townhouses in SDC A and B from the seismic provisions of the IRC. In these cases, only the wind bracing requirements must be met.

As illustrated in **FIGURE 6.6**, the information in the seismic bracing table is based on:

- Soil site classification D (see *Soil Site Classes* on page 133)

- 10-foot wall height

- 10 psf floor dead load

- 15 psf roof and ceiling dead load

- Braced wall line spacing of 25 feet or less

FIGURE 6.6
Basis for seismic bracing table

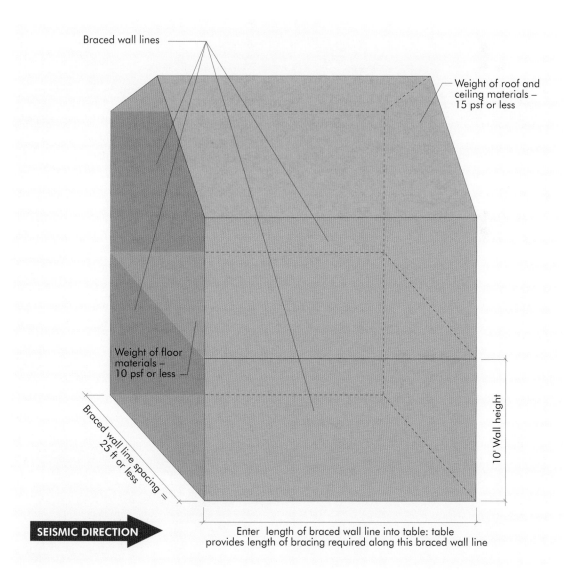

Braced wall lines

Weight of roof and ceiling materials – 15 psf or less

Weight of floor materials – 10 psf or less

Braced wall line spacing = 25 ft or less

10' Wall height

SEISMIC DIRECTION

Enter length of braced wall line into table: table provides length of bracing required along this braced wall line

The footnotes to the seismic bracing table include adjustment factors to accommodate variations from these assumptions for the wide range of residential structures covered by the IRC. These adjustment factors are included in IRC Table R602.10.1.2(3), reproduced in **TABLE 6.3**. While not all of these adjustments are new to the IRC, many have been taken from other sections and collected into a common location in the 2009 edition.

TABLE 6.2

Bracing requirements based on Seismic Design Category (a function of braced wall line length)

IRC Table R602.10.1.2(2)[a,b,c]

Soil Class D[a] Wall Height = 10 ft 10 PSF Floor Dead Load 15 PSF Roof/Ceiling Dead Load Braced Wall Line Spacing ≤ 25 ft			Minimum Total Length (Feet) of Braced Wall Panels Required Along Each Braced Wall Line			
Seismic Design Category (SDC)	Story Location	Braced Wall Line Length	Method LIB	Methods DWB, SFB, GB, PBS, PCP, HPS	Method WSP	Cont. Sheathing
SDC A and B, and Detached Dwellings in C			Exempt from Seismic Requirements Use Table R602.10.1.2(1) for bracing requirements			
		10	2.5	2.5	1.6	1.4
		20	5.0	5.0	3.2	2.7
		30	7.5	7.5	4.8	4.1
		40	10.0	10.0	6.4	5.4
		50	12.5	12.5	8.0	6.8
SDC C		10	NP	4.5	3.0	2.6
		20	NP	9.0	6.0	5.1
		30	NP	13.5	9.0	7.7
		40	NP	18.0	12.0	10.2
		50	NP	22.5	15.0	12.8
		10	NP	6.0	4.5	3.8
		20	NP	12.0	9.0	7.7
		30	NP	18.0	13.5	11.5
		40	NP	24.0	18.0	15.3
		50	NP	30.0	22.5	19.1
		10	NP	3.0	2.0	1.7
		20	NP	6.0	4.0	3.4
		30	NP	9.0	6.0	5.1
		40	NP	12.0	8.0	6.8
		50	NP	15.0	10.0	8.5
SDC D₀ or D₁		10	NP	6.0	4.5	3.8
		20	NP	12.0	9.0	7.7
		30	NP	18.0	13.5	11.5
		40	NP	24.0	18.0	15.3
		50	NP	30.0	22.5	19.1
		10	NP	8.5	6.0	5.1
		20	NP	17.0	12.0	10.2
		30	NP	25.5	18.0	15.3
		40	NP	34.0	24.0	20.4
		50	NP	42.5	30.0	25.5

TABLE 6.2 (CONT.)

**Bracing
requirements
based on
Seismic Design
Category (a
function of
braced wall
line length)**

*IRC Table
R602.10.1.2(2)[a,b,c]*

Soil Class D[a] Wall Height = 10 ft 10 PSF Floor Dead Load 15 PSF Roof/Ceiling Dead Load Braced Wall Line Spacing ≤ 25 ft			Minimum Total Length (Feet) Of Braced Wall Panels Required Along Each Braced Wall Line			
Seismic Design Category (SDC)	Story Location	Braced Wall Line Length	Method LIB	Methods DWB, SFB, GB, PBS, PCP, HPS	Method WSP	Cont. Sheathing
SDC A and B, and Detached Dwellings in C			**Exempt from Seismic Requirements Use Table R602.10.1.2(1) for bracing requirements**			
SDC D₂	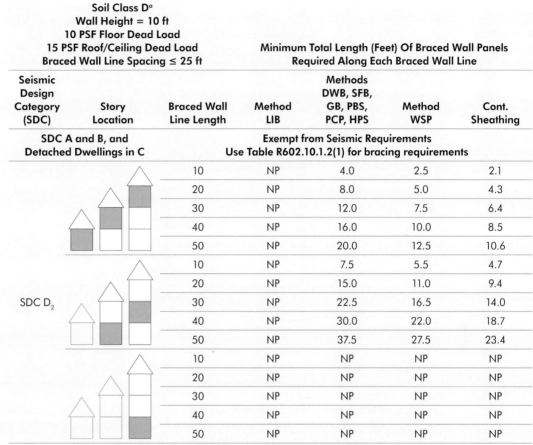	10	NP	4.0	2.5	2.1
		20	NP	8.0	5.0	4.3
		30	NP	12.0	7.5	6.4
		40	NP	16.0	10.0	8.5
		50	NP	20.0	12.5	10.6
		10	NP	7.5	5.5	4.7
		20	NP	15.0	11.0	9.4
		30	NP	22.5	16.5	14.0
		40	NP	30.0	22.0	18.7
		50	NP	37.5	27.5	23.4
		10	NP	NP	NP	NP
		20	NP	NP	NP	NP
		30	NP	NP	NP	NP
		40	NP	NP	NP	NP
		50	NP	NP	NP	NP

a. Wall bracing lengths are based on a soil site class "D." Interpolation of bracing length between the S_{ds} values associated with the Seismic Design Categories shall be permitted when a site-specific S_{ds} value is determined in accordance with Section 1613.5 of the International Building Code.
b. Foundation cripple wall panels shall be braced in accordance with Section R602.10.9.
c. Methods of bracing shall be as described in Sections R602.10.2, R602.10.4 and R602.10.5.

Footnote a of IRC Table R602.10.1.2(2) warrants some explanation. Soil site class D represents a fairly soft soil (see "Soil Site Classes" on page 133). When the soil properties are not known in sufficient detail to determine a specific soil site class, site Class D shall be used unless the building official or geotechnical data determines that site class E or F are present. If site class E or F are present, IBC Section 1613.5 provides a methodology for determining the site S_{DS} and converting it into a SDC that can be used directly in IRC Table R602.10.1.2(2).

TABLE 6.3

Adjustment factors to the length of required seismic wall bracing^a

IRC Table R602.10.1.2(3)

Adjustment Based On:		Multiply Length of Bracing per Wall Line by:		Applies to:
Story height^b (Section R301.3)		≤ 10 ft	1.0	
		> 10 ft ≤ 12 ft	1.2	
Braced wall line spacing, townhouses in SDC A–C^{b,c}		≤ 35 ft	1.0	All bracing methods - Sections R602.10.2, R602.10.4 and R602.10.5
		> 35 ≤ 50 ft	1.43	
Wall dead load		> 8 ≤ 15	1.0	
		≤ 8 psf	0.85	
Roof/ceiling dead load for wall supporting^b	Roof only or roof plus one story	≤ 15 psf	1.0	
	Roof only	> 15 psf ≤ 25 psf	1.1	
	Roof plus one story	> 15 psf ≤ 25 psf	1.2	
Walls with stone or masonry veneer in SDC C–D₂		See Section R703.7		
Cripple walls		See Section R602.10.9		

a. The total length of bracing required for a given wall line is the product of all applicable adjustment factors.
b. Linear interpolation shall be permitted.
c. Braced wall line spacing and adjustments to bracing length in SDC D_0 to D_2 shall comply with Section R602.10.1.5.

Each of the adjustments described in IRC Table R602.10.1.2(3) are explained below:

- **Story height (IRC Section R301.3)** – The amount of bracing required for seismic loads is directly related to the mass/weight of the structure. As a wall gets taller, its mass increases, thus requiring more bracing to resist the resulting seismic loads.

- **Braced wall line spacing, townhouses in SDC A-C** – This adjustment factor actually only applies to townhomes in SDC C because of the seismic exception discussed previously (IRC Section R301.2.2: detached one- and two-family dwellings in SDC A, B and C and townhouses in SDC A and B are exempt from the seismic requirements of the code). Townhomes in SDC C are permitted to have a 35-foot to 50-foot braced wall line spacing using this adjustment factor. The 25-foot wall line spacing (first column heading IRC Table R602.10.1.2(2)) applies to one- and two-family dwellings in SDC D_0, D_1 and D_2 and townhouses in SDC C, D_0, D_1 and D_2. However, for one room only per dwelling unit, the 25-foot braced wall line spacing for these structures can be increased to a maximum of 35 feet using IRC Section R602.10.1.5 and the adjustment factors in IRC Table R602.10.1.5.

- **Wall dead load** – IRC Table R602.10.1.2(2) is based on a wall weight of 8 to 15 psf. If a lighter weight wall is used, a reduction is permitted in the amount of bracing required. (Again, more mass requires more bracing and less mass requires less bracing.) Note that a standard wood-framed stud wall has a weight of 11 to 12 psf (Table C3-1 of ASCE 7-05).

- **Roof/ceiling dead load for wall supporting** – IRC Table R602.10.1.2(2) is based on a roof/ceiling weight of 15 psf or less. As the roof weight/mass increases, so does the amount of bracing required. Note that a standard wood-framed roof ceiling with lightweight asphalt shingles or wood shingles has a weight that varies between 10 and 15 psf (Table C3-1 of ASCE 7-05).

- **Walls with stone or masonry veneer in SDC C, D_0, D_1 and D_2** – This is another case of more mass requiring more bracing. Note that two tables are referenced, one for low seismic (SDC A, B and C) and one for high seismic. Once again, detached one- and two-family dwellings in SDC A, B and C and townhouses in SDC A and B are exempt from the seismic requirements of the code (IRC Section R301.2.2). As such, the low seismic table (IRC Table R602.12(1)) is applicable for townhouses only. In converse fashion, the high seismic table (IRC Table R602.12(2)) appears only to apply to detached one- and two-family dwellings and not to townhouses.

- **Cripple walls** – The reference to IRC Section R602.10.9 provides the user with adjustments for cripple wall bracing. This is addressed in **CHAPTER 9** and has not changed this code cycle.

SOIL SITE CLASSES

The seismic bracing requirements are based on soil site class D. Site classifications (discussed in the IBC, Section 1613.5.2) range from A to F. Soil site class A represents hard rock and site class F represents soils generally not considered ideal for construction. Site class D is the default soil condition and can be used for all cases, unless the building official or geotechnical survey determines that site class E or F are present. Site class E and F are very soft, loamy, sandy soils, and soils with soft and medium clays. These site classes are undesirable for prescriptive applications because softer soils amplify the earthquake ground motion. The local building official will know if such soils are present in a given area and will be familiar with the necessary requirements for building in such conditions. Contact your local building official for guidance – they issue the building permits.

INTERPOLATION

Interpolation is a mathematical tool that can be used to determine a value from a table that does not include the value you are seeking, but lies between values that are given in the table. For example, if you are looking for the amount of bracing required for a building sited where the basic wind speed is 110 mph with a braced wall line spacing of 23 feet, you will discover in IRC Table R602.10.1.2(1) that there are values given for braced wall line spacings of 20 and 30 feet, but not for 23 feet. In such cases, the user can assume the next larger braced wall line spacing (30 feet, in this example), but doing so may result in more bracing than is technically required, which may ultimately restrict the use of desired architectural features. A better solution is to use interpolation to find the value between 20 feet and 30 feet that corresponds to 23 feet.

Although the equation for interpolation (Equation 1) may appear daunting, the correct value is simple to compute with a standard calculator and, with a little practice, can actually be performed quite quickly. If interpolation is used, footnotes and adjustment factors are initially disregarded and then applied to the value after Equation 1 has been solved.

$$y = y_a + \frac{(x-x_a)(y_b-y_a)}{(x_b-x_a)}$$ Equation 1 (See Step 2 for an explanation of terms)

Example: We are looking for the amount of wind bracing required for the first story of two stories in a 110 mph wind zone. We would like to use gypsum board double-sided bracing and our braced wall line spacing is 23 feet. Here are the steps to interpolate the correct value:

Step 1. Find the proper table and look for a length of bracing required for a braced wall line spacing of 23 feet.

				Minimum total length (feet) of braced wall panels required along each braced wall line		
Basic Wind Speed	**Story Location**	**Braced Wall Line Spacing (ft)**	**Method LIB[f,h]**	**Method GB (Double-Sided)[g]**	**Methods DWB, WSP, SFB, PBS, PCP, HPS[f,i]**	**Cont. Sheathing**
< 110 Mph		10	10.5	10.5	6.0	5.0
		20	19.0	19.0	11.0	9.5
		30	27.5	27.5	16.0	13.5
		40	36.0	36.0	20.5	17.5
		50	44.0	44.0	25.5	21.5
		60	52.5	52.5	30.0	25.5

Exposure Category B
30 ft mean roof height
10 ft eave to ridge height
10 ft wall height
2 braced wall line

The highlighted cells from this excerpt of IRC Table R602.10.1.2(1) reveal that a value for 23 feet is not offered; however, values are given for 20 and 30 feet. These are the values we need to interpolate.

Step 2. Create a table with the values that we do know and the value that we want to know, as shown below:

Braced Wall Line Spacing (ft)	Method GB (Double-Sided)[g]
20	19
23	?
30	27.5

The values of this table are represented as:

$x_a = 20$	$y_a = 19$
$x = 23$	$y = ?$
$x_b = 30$	$y_b = 27.5$

Step 3. Substitute the values into the interpolation equation:

$$y = y_a + \frac{(x - x_a)(y_b - y_a)}{(x_b - x_a)} \qquad \text{Equation 1}$$

$$y = 19 + \frac{(23-20)(27.5-19)}{(30-20)}$$

$$y = 21.6 \text{ feet of bracing required}$$

As stated previously, the alternative is to use the next larger braced wall line spacing: 30 feet, in this example, which would require 27.5 feet of bracing. Using interpolation, we reduce the amount of bracing required (27.5 feet – 21.6 feet = 5.9 feet) by nearly 6 feet! 6 feet of reduced bracing may be well worth the few minutes it takes to interpolate.

The above example is for wind bracing. The same principles apply to seismic bracing, but interpolation is applied to the braced wall line length instead of braced wall line spacing.

Length requirement for continuous sheathing bracing methods

The length requirement for continuous sheathing is covered here because of a number of changes in the 2009 IRC that impact the method. Details on the continuous sheathing methods themselves can be found in **CHAPTER 5**.

In the past, the bracing length adjustment for Method CS-WSP (continuous wood structural panel sheathing) was calculated based on the size of the wall openings (door or window) in the braced wall line. The ICC Ad Hoc Wall Bracing Committee determined that the differences in wall opening sizes were of little enough impact that a single factor could be applied, regardless of the opening size. A continuous sheathing column was subsequently added to the wind and seismic bracing tables to eliminate the need for an additional calculation that factored in opening size.

In addition, a new continuous sheathing bracing method was added: Method CS-SFB (continuous structural fiberboard sheathing). While the lengths of bracing required are equal for both of these methods, Method CS-SFB includes a number of limitations. For this reason, Method CS-SFB is addressed in a different section of the Code (IRC Section R602.10.5) from Method CS-WSP (IRC Section R602.10.4). Both are detailed in **CHAPTER 5**.

Examples: Determining length of bracing

Example 6.1:

Determining length of bracing using intermittent Method SFB (structural fiberboard sheathing).

Given:

- The house is in SDC A with a 105 mph Exposure B design wind speed.

- 48-inch Method SFB (structural fiberboard sheathing) bracing is used.

- Braced wall line has no stories above it.

- The distance between braced wall lines is 30 feet.

- The garage is 30 feet deep.

- Roof eave-to-ridge height is 10 feet.

- *FIGURE 6.7*.

FIGURE 6.7

Example using Method SFB (structural fiberboard sheathing)

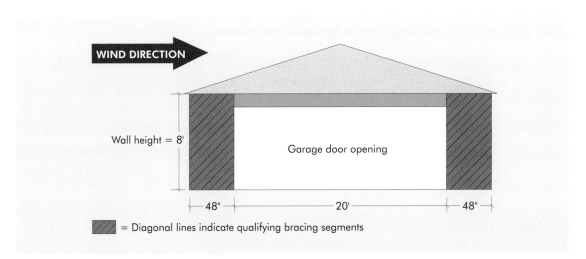

Solution:

Step 1. Determine which bracing tables, wind and/or seismic, are required for analysis.

The wall line is assumed to be a part of a detached one- or two-family residence. In accordance with IRC Section R301.2.2, it is exempt from seismic design (it is located in SDC A) so only the wind bracing tables apply.

Step 2. Determine how much wind bracing is required for the given scenario.

- From IRC Table R602.10.1.2(1) (**TABLE 6.1**), we can see that for a wall line in a single-story structure, Method SFB bracing, wind speeds less than 110 mph (Exposure B), and a braced wall line spacing of 30 feet: <u>8.5 feet of bracing is required</u> and panels shall not be spaced greater than 25 feet on center.

- Adjust values:

  - In accordance with Footnote d of IRC Table R602.10.1.2(1) (**TABLE 6.1**), use of a wall height of 8 feet permits a reduction in bracing length by multiplying length by 0.9.

    8.5 feet x 0.90 = 7.65 feet of bracing required

Step 3. Determine how much qualified bracing is present in the braced wall line.

$$\frac{(48 \text{ inches } + 48 \text{ inches})}{12 \text{ inches per foot}} = 8 \text{ feet of bracing available}$$

Step 4. This braced wall line meets the minimum bracing requirement for the given conditions: the 8 feet of bracing present is more than the 7.65 feet required.

Example 6.2:

Determining length of bracing using Method PBS (particleboard sheathing) in a 10-foot tall wall.

Given:

- The structure is a single-family residence in SDC D_0 with a 90 mph Exposure B design wind speed.

- 48-inch Method PBS (particleboard sheathing) bracing is used.

- Braced wall line has one story above it.

- Roof eave-to-ridge height is 7 feet.

- Light-weight wood construction with asphalt shingles.

- The length of the wall being braced is 30 feet.

- The distance between braced wall lines is 30 feet.

- **FIGURE 6.8**.

FIGURE 6.8

Example using Method PBS (particleboard sheathing)

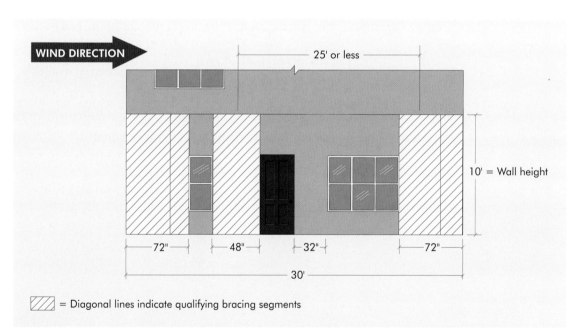

= Diagonal lines indicate qualifying bracing segments

Solution:

Step 1. Determine which bracing tables, wind and/or seismic, are required for analysis.

The wall line is assumed to be a part of a detached one- or two-family residence. In accordance with IRC Section R301.2.2, this structure is located in SDC D_0 and therefore not exempt from seismic design. Both wind and seismic loading must be considered.

Step 2. Determine how much wind bracing is required for given scenario.

- From IRC Table R602.10.1.2(1) (**TABLE 6.1**), we can see that for a first-story wall line of a two-story structure, 10-foot tall wall, Method PBS bracing, 90 mph winds (Exposure B), and a braced wall line spacing of 30 feet: <u>10.5 feet of bracing is required</u> and panels shall not be spaced greater than 25 feet on center.

- Adjust values:

  ○ In accordance with Footnote c of IRC Table R602.10.1.2(1) (**TABLE 6.1**), for a two-story building with a 7-foot ridge height, multiply the required bracing length by 0.91.

    10.5 feet x 0.91 = 9.56 feet of wind bracing required

Step 3. Determine how much seismic bracing is required.

- From IRC Table R602.10.1.2(2) (**TABLE 6.2**), for a first-story wall line of a two-story structure, 10-foot tall wall, Method PBS bracing, SDC D_0, a braced wall line spacing of 30 feet, and a braced wall line length of 30 feet: <u>18 feet of bracing is required</u>.

- Adjust values:

  ○ For braced wall line spacing of 30 feet and a braced wall line length of 30 feet, the room bordered by these braced wall lines is 900 square feet. As such, in accordance with IRC Table R602.10.1.5, an adjustment factor of 1.2 is required.

    18 feet x 1.2 = 21.6 feet of seismic bracing is required

Step 4. Determine how much qualified bracing is present in the braced wall line.

$$\frac{(72 \text{ inches} + 72 \text{ inches} + 48 \text{ inches})}{12 \text{ inches per foot}} = 16 \text{ feet of bracing available}$$

Step 5. Of the two determined bracing wall lengths, the seismic requirement is greater and controls at 21.6 feet; however, the braced wall line does not meet this requirement for the given conditions. The 16 feet of bracing present is less than the 21.6 feet required. Possible solutions include:

- Add additional bracing panel length (an unlikely solution as the braced wall line length is 30 feet long and 21.6 feet is required).

- For seismic applications, IRC Section R602.10.1.5 provides for braced wall lines to be no farther apart than 35 feet for a single room – not to exceed 900 square feet – in each dwelling unit, as long as all of the other braced wall lines in that unit are spaced at 25 feet or less. As a solution for this example, ensure that other braced wall lines in the structure do not exceed 25 feet on center spacing, then the 1.2 multiplier applied to the required amount of bracing need not be applied to the amount of seismic bracing required. As such, 18 feet of bracing are required. Because only 16 feet of bracing are available, this alone is not an effective solution; however, it may be joined with Method WSP (requires 13.5 feet of bracing) or Method CS-WSP or CS-SFB (requires 11.5 feet of bracing). All three of these methods provide sufficient bracing, assuming the 900 square foot provisions have been met.

Example 6.3:

Determining length of bracing using Method HPS (hardboard panel siding), away from corner, in 9-foot tall wall with offsets in braced wall line.

Given:

- The house is in SDC A with a 90 mph Exposure B design wind speed.

- 48-inch-wide Method HPS (hardboard panel siding) bracing is used.

- Braced wall line has one story above it.

- Lightweight wood construction with asphalt shingles.

- Roof eave-to-ridge height is 9 feet.

- The distance between braced wall lines is 25 feet.

- ***FIGURE 6.9***.

FIGURE 6.9

Example using Method HPS (hardboard panel siding)

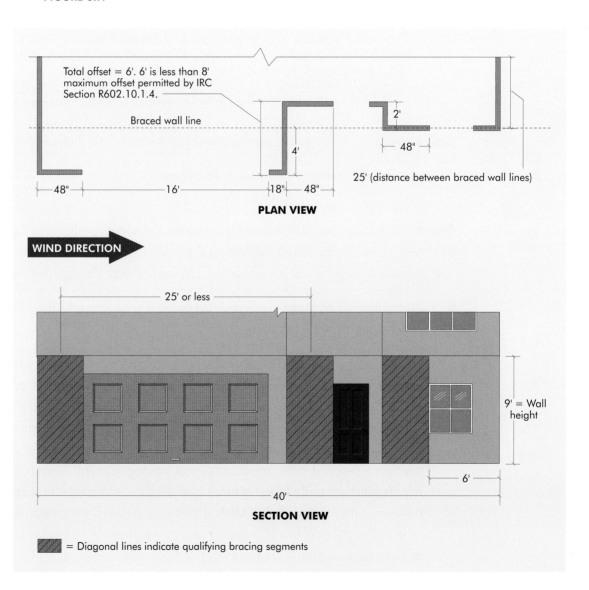

Total offset = 6'. 6' is less than 8' maximum offset permitted by IRC Section R602.10.1.4.

Braced wall line

2'

48"

4'

25' (distance between braced wall lines)

48"   16'   18"   48"

PLAN VIEW

WIND DIRECTION

25' or less

9' = Wall height

6'

40'

SECTION VIEW

= Diagonal lines indicate qualifying bracing segments

Solution:

Step 1. Determine which bracing tables, wind and/or seismic, are required for analysis.

The wall line is assumed to be a part of a detached one-or two-family residence. In accordance with IRC Section R301.2.2, it is exempt from seismic design (it is located in SDC A) so only the wind bracing tables apply.

Step 2. Determine how much wind bracing is required.

- From IRC Table R602.10.1.2(1) (**TABLE 6.1**), we can see that for a first-story wall line of a two-story structure, 9-foot tall wall, Method HPS bracing, 90 mph winds (Exposure B), and a braced wall line spacing of 25 feet: <u>9 feet of bracing is required</u> by interpolation (see page 134) (7.5 feet required for 20 foot spacing and 10.5 feet required for 30 foot spacing; interpolation yields 9 feet).

- Adjust values:

    - In accordance with Footnote d of IRC Table R602.10.1.2(1) (**TABLE 6.1**), use of 9-foot tall walls permits an adjustment of 0.95.

        9 feet x 0.95 = 8.6 feet of bracing required

    - In accordance with Footnote c of IRC Table R602.10.1.2(1) (**TABLE 6.1**), an eave-to-ridge height of 9 feet permits an adjustment of 0.97, by interpolation.

        8.6 feet x 0.97 = 8.34 feet of bracing required

Step 3. Determine how much qualified bracing is present in the braced wall line.

Note that all offsets are within 4 feet of the effective (imaginary) braced wall line, so all qualified full-width panels on all three wall sections count towards bracing the braced wall line.

$$\frac{(48 \text{ inches} + 48 \text{ inches} + 48 \text{ inches})}{12 \text{ inches per foot}} = 12 \text{ feet of bracing available}$$

Step 4. This braced wall line meets the minimum bracing requirements for the given conditions: the 12 feet of bracing present is more than the 8.34 feet required. The existing bracing is sufficient.

Note that the panel on the right is offset 6 feet from one end of the braced wall line. This meets both the IRC Section R602.10.1.4 offset rules (12-feet 6-inch maximum offset of a single braced wall panel from one end of a braced wall line) and the 12 feet 6 inch cumulative total of offsets from each end. This is discussed in **CHAPTER 7**.

Example 6.4:

Determining length of bracing using Method PCP (portland cement plaster).

Given:

- The house is in SDC A with a 105 mph Exposure C design wind speed.

- Method PCP (portland cement plaster) bracing is used. Method PCP adds 10.4 psf to the weight of the wall. The weight associated with the plaster does not require an adjustment factor within SDC A, as one- and two-story residences are exempt from the seismic requirements of the IRC.

- Light-weight roof construction with asphalt shingles.

- Roof eave-to-ridge height is 7 feet.

- Braced wall line has one story above it.

- The distance between braced wall lines is 40 feet.

- ***FIGURE 6.10.***

FIGURE 6.10

Example using Method PCP (portland cement plaster)

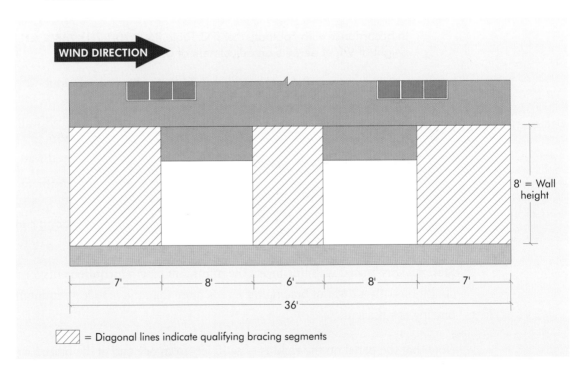

Solution:

Step 1. Determine which bracing tables, wind and/or seismic, are required for analysis.

As the wall line is located in a SDC A, all structures covered by the IRC are exempt from seismic design. Only wind must be considered.

Step 2. Determine how much wind bracing is required.

- From IRC Table R602.10.1.2(1) (***TABLE 6.1***), we can see that for a first-story wall line of a two-story structure, Method PCP bracing, 105 mph winds (Exposure C), and a braced wall line spacing of 40 feet: If the 110 mph portion of IRC Table R602.10.1.2(1) is used, <u>20.5 feet of bracing is required</u>. If, however, interpolation is used between the 100 mph and the 110 mph portions of the table, only 18.75 feet of bracing is required. For this example, interpolation will not be used, even though it provides a workable solution. Instead, this example illustrates the addition of an interior braced wall line in the solution.

- Adjust values:
  - In accordance with Footnote b of IRC Table R602.10.1.2(1) (***TABLE 6.1***), a two-story Exposure C building requires an adjustment of 1.3.

    20.5 feet x 1.3 = 26.65 feet of bracing required

  - In accordance with Footnote c of IRC Table R602.10.1.2(1) (***TABLE 6.1***), for eave-to-ridge height of 7 feet, by interpolation:

    26.65 feet x 0.91 = 24.25 feet of bracing required

  - In accordance with Footnote d of IRC Table R602.10.1.2(1) (***TABLE 6.1***), for an 8-foot maximum wall height:

    24.25 feet x 0.90 = 21.83 feet of bracing required

Step 3. Determine how much qualified bracing is present in the braced wall line.

(7 feet + 6 feet + 7 feet) = 20 feet of bracing available

Step 4. This braced wall line does not meet the minimum bracing requirement for the given conditions: the 20 feet of bracing present is less than the 21.83 feet required. One possible solution is to add an interior braced wall line as shown in **FIGURE 6.11**. This results in braced wall line spacings of 25 and 15 feet. In accordance with IRC Figure R602.10.1.4(4), the greater of the two spacings (25 feet) controls. As step 4 does not provide a workable solution, repeat steps 2-4.

FIGURE 6.11

Adding a braced wall line to the interior of the structure to reduce bracing on exterior wall lines

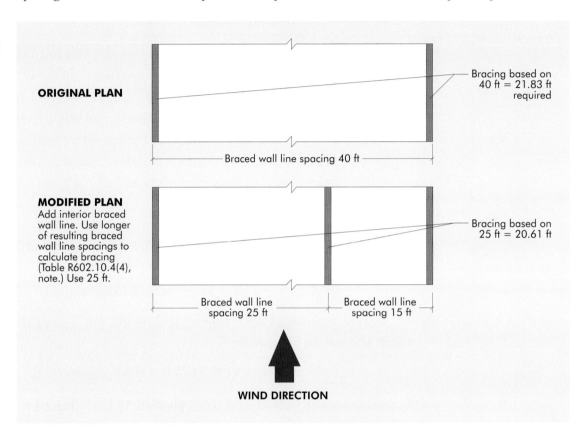

ORIGINAL PLAN

Bracing based on 40 ft = 21.83 ft required

— Braced wall line spacing 40 ft —

MODIFIED PLAN
Add interior braced wall line. Use longer of resulting braced wall line spacings to calculate bracing (Table R602.10.4(4), note.) Use 25 ft.

Bracing based on 25 ft = 20.61 ft

Braced wall line spacing 25 ft

Braced wall line spacing 15 ft

WIND DIRECTION

Step 2 (Repeat). Determine how much wind bracing is required.

- From IRC Table R602.10.1.2(1) (**TABLE 6.1**), we can see that for a first-story wall line of a two-story structure, Method PCP bracing, 105 mph winds, and a braced wall line spacing of 25 feet: 13.5 feet of bracing is required by interpolation (see page 134) (11 feet required for 20 foot spacing and 16 feet required for 30 foot spacing; interpolation yields 13.5 feet). Note that the next largest wind speed –100 mph – was used for this example. Interpolation between 100 mph and 110 mph could have been used for a reduced braced wall line requirement.

- Adjust values:

  - In accordance with Footnote b of IRC Table R602.10.1.2(1) (*TABLE 6.1*), a two-story Exposure C building requires an adjustment of 1.3.

    13.5 feet x 1.3 = 17.6 feet of bracing required

  - In accordance with Footnote c of IRC Table R602.10.1.2(1) (*TABLE 6.1*), for an eave-to-ridge height of 7 feet, by interpolation:

    17.6 feet x 0.91 = 16 feet of bracing required

  - In accordance with Footnote d of IRC Table R602.10.1.2(1) (*TABLE 6.1*), for an 8 foot maximum wall height:

    16 feet x 0.90 = 14.4 feet of bracing required

  - In accordance with Footnote e of IRC Table R602.10.1.2(1) (*TABLE 6.1*), three braced wall lines require an additional adjustment of 1.3.

    14.4 feet x 1.3 = 18.72 feet of bracing required

Step 3 (Repeat). See Step 3 on page 143.

Step 4 (Repeat). This braced wall line meets the minimum bracing requirement for the given conditions: the 20 feet of bracing present is more than the 18.72 feet required.

Note that once the braced wall line length requirement has been met by the braced wall length provided, it is unnecessary to seek additional reductions to the required braced wall line length.

Examples: Determining length of bracing when using narrow wall (less than 48 inches) braced wall panels

Example 6.5:

Method CS-G (wood structural panel adjacent to garage door openings and supporting roof loads only)

Method CS-G permits bracing panels on either side of single-story garages that have a 4:1 height-to-length ratio – tabular values based on wall height provided in IRC Table R602.10.4.2 (*TABLE 5.6*) – provided that the roof coverings do not exceed 3 psf. This bracing option is included in IRC Table R602.10.4.1 and can only be used when the wall line is continuously sheathed with wood structural panels (Method CS-WSP) in accordance with IRC Section R602.10.4. Note that the 3 psf limitation on roof mass has been determined to be an issue only when selecting seismic bracing. The limitation does not apply to the selection of wind bracing.

Given:

- The single-family garage is located in SDC C with a 90 mph Exposure B design wind speed.

- Method CS-WSP bracing is used per IRC Section R602.10.4 (see also IRC Table R602.10.4.1).

- The braced wall line has a roof with a dead load of 15 psf. (Note that Footnote b of IRC Table R602.10.4.1 is a seismic requirement and does not apply in this case (per IRC Section R301.2.2) because the structure is a garage located in SDC C). An increase in roof mass increases seismic loads only.

- The distance between braced wall lines is 25 feet.

- Roof eave-to-ridge height is 6 feet.

- *FIGURE 6.12*.

FIGURE 6.12

Example using Method CS-G (wood structural panel adjacent to garage door openings and supporting roof loads only)

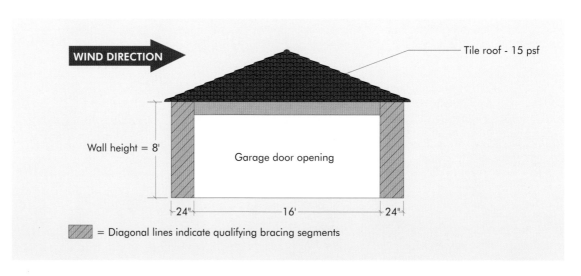

Solution:

Step 1. Determine which bracing tables, wind and/or seismic, are required for analysis.

As the wall line is a part of a single-family residence in SDC C, it is exempt from seismic design. Only wind must be considered.

Step 2. Determine how much wind bracing is required.

- From IRC Table R602.10.1.2(1) (**TABLE 6.1**), we can see that for continuous sheathing, braced wall line spacing of 25 feet, and wind speeds of 90 mph or less (Exposure B): <u>4.25 feet of bracing is required</u> by interpolation (see page 134) (3.5 feet required for 20 foot spacing and 5 feet required for 30 foot spacing; interpolation yields 4.25 feet).

- Adjust values:

  - In accordance with Footnote c of IRC Table R602.10.1.2(1) (**TABLE 6.1**), a single-story building with a 6-foot eave-to-ridge height is permitted an adjustment of 0.76 by interpolation.

    4.25 feet x 0.76 = 3.23 feet of bracing required

  - In accordance with Footnote d of IRC Table R602.10.1.2(1) (**TABLE 6.1**), a building with an 8-foot wall height is permitted an adjustment of 0.90.

    3.23 feet x 0.90 = 2.91 feet of bracing required

Step 3. Determine how much qualified bracing is present in the braced wall line.

24 inches + 24 inches = 48 inches = 4 feet of bracing available

Step 4. This braced wall line meets the minimum bracing requirement for the given conditions: the 4 feet of bracing present is more than the 2.91 feet required. Note that IRC Section R602.10.1.2 requires a minimum bracing requirement after all adjustments of 48 inches for each braced wall line.

Example 6.6:

Method CS-SFB (continuous structural fiberboard sheathing)

Given:

- The house is in SDC B with a 100 mph Exposure B design wind speed.

- Method CS-SFB bracing is used per IRC Section R602.10.5 (see also IRC Table R602.10.5.2)

- Braced wall line has one story above it.

- The distance between braced wall lines is 20 feet.

- Roof eave-to-ridge height is 5 feet.

- ***FIGURE 6.13***.

FIGURE 6.13

Example using Method CS-SFB (continuous structural fiberboard sheathing)

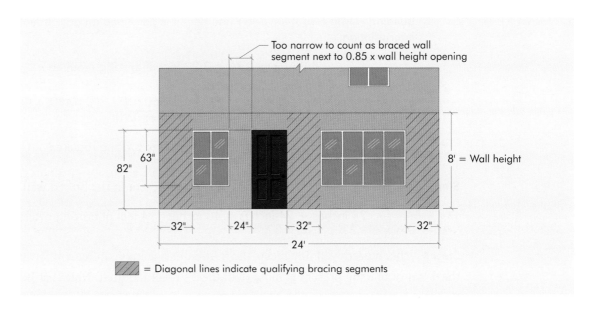

Solution:

Step 1. Determine which bracing tables, wind and/or seismic, are required for analysis.

As the wall line is a part of a single-family residence in SDC B, it is exempt from seismic design. Only wind must be considered.

Step 2. Determine how much wind bracing is required.

- From IRC Table R602.10.1.2(1) (***TABLE 6.1***), we can see that for a first-story wall line in a two-story residence, a 100 mph basic wind speed (Exposure B), continuously sheathed with a wall bracing spacing of 20 feet: <u>8 feet of bracing is required</u>.

- Adjust values:

    - In accordance with Footnote c of IRC Table R602.10.1.2(1) (***TABLE 6.1***), for an eave-to-ridge height of 5 feet:

    $$8 \text{ feet x } 0.85 = 6.8 \text{ feet of bracing required}$$

    - In accordance with Footnote d of IRC Table R602.10.1.2(1) (***TABLE 6.1***), for a wall height of 8 feet:

    $$6.8 \text{ feet x } 0.9 = 6.12 \text{ feet of bracing required}$$

Step 3. Determine how much qualified bracing is present in the braced wall line.

Check full-height segment lengths to ensure that they meet minimum length requirements (per IRC Table R602.10.4.2) for the size of the window or door opening adjacent. Note that the 24-inch wall segment adjacent to the door is too narrow to count. A braced wall panel adjacent to an 82-inch clear height opening requires a minimum length of 32 inches.

$$\frac{(32 \text{ inches} + 32 \text{ inches} + 32 \text{ inches})}{12 \text{ inches per foot}} = 8 \text{ feet of bracing available}$$

Step 4. This braced wall line meets the minimum bracing requirement for the given conditions: the 8 feet of bracing present is more than the 6.12 feet required.

Note that neither of the adjustments to the required bracing length are necessary, as it was determined that 8 feet of bracing is required, and the braced wall length requirement from IRC Table R602.10.1.2(1) is also 8 feet. The user must be careful, however, when ignoring the braced wall line length adjustments in the footnotes if the amount of bracing appears to be sufficient without adjustment. Some of the adjustments increase the amount of bracing required. For example, the eave-to-ridge adjustment in Footnote c of IRC Table R602.10.1.2(1): when the eave-to-ridge height is 10 feet or greater, the required bracing must be increased (to compensate for the increased surface area of the structure resisting the wind load).

Example 6.7:

Method CS-PF (continuous portal frame)

Given:

- The structure is a single-family residence in SDC B with an 85 mph Exposure B design wind speed.
- 10-foot wall height.
- Method CS-PF bracing is used per IRC Section R602.10.4 (see also IRC Table R602.10.4.1.1).
- Braced wall line has one story above it.
- The distance between braced wall lines is 20 feet.
- Roof eave-to-ridge height is 6 feet.
- ***FIGURE 6.14***.

FIGURE 6.14

Example using Method CS-PF (continuous portal frame) at garage wall

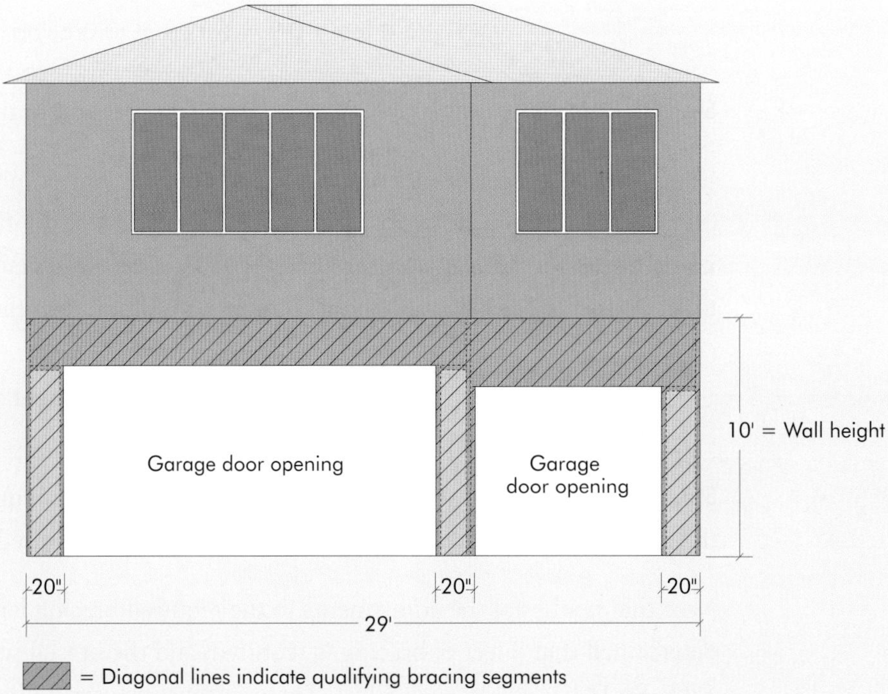

Garage door opening
Garage door opening
10' = Wall height
20"
20"
20"
29'
= Diagonal lines indicate qualifying bracing segments

Solution:

Step 1. Determine which bracing tables, wind and/or seismic, are required for analysis.

As the wall line is a part of a single-family residence in SDC B, it is exempt from seismic design. Only wind must be considered.

Step 2. Determine how much wind bracing is required.

- From IRC Table R602.10.1.2(1) (**TABLE 6.1**), we can see that for a first-story wall line of a two-story structure, 10-foot tall wall, 85 mph Exposure B wind zone, Method CS-PF bracing, and a braced wall line spacing of 20 feet: <u>5.5 feet of bracing is required</u>. Note also that a qualified return corner (IRC Section R602.10.4.4(1)) is also required at each end of the braced wall line. In lieu of a return corner, a minimum 800-lb capacity hold down must be attached to the corner studs.

- Adjust values:

  - In accordance with Footnote c of IRC Table R602.10.1.2(1) (**TABLE 6.1**), for a two-story building with an a eave-to-ridge height of 6 feet, by interpolation:

$$5.5 \text{ feet} \times 0.88 = 4.84 \text{ feet of bracing required}$$

Step 3. Check minimum length requirement for Method CS-PF.

Per IRC Table R602.10.4.2, for a 10-foot tall wall, the minimum length permitted for this segment is 20 inches. (See **CHAPTER 5** for more information on Method CS-PF minimum length.)

Step 4. Determine how much qualified bracing is present in the braced wall line.

$$\frac{(20 \text{ inches} + 20 \text{ inches} + 20 \text{ inches})}{12 \text{ inches per foot}} = 5 \text{ feet of bracing available}$$

Step 5. This braced wall line meets the minimum bracing requirement for the given conditions: the 5 feet of bracing present is more than the 4.84 feet feet required.

Example 6.8:

Method CS-PF (continuous portal frame) and Method CS-WSP (continuous wood structural panel sheathing) with offsets in braced wall line for a single-story structure.

Given:

- The structure is a one-story, single-family residence in SDC C with a 90 mph Exposure B design wind speed.

- 9-foot wall height.

- Method CS-PF bracing is used, per IRC Section R602.10.4.1.1 (see also IRC Tables R602.10.4.1.1 and R602.10.4.2).

- Braced wall line supports a roof only.

- The distance between braced wall lines is 30 feet.

- Roof eave-to-ridge height is 7.5 feet.

- *FIGURE 6.15*.

FIGURE 6.15

Example using Method CS-PF (continuous portal frame) at offset wall line

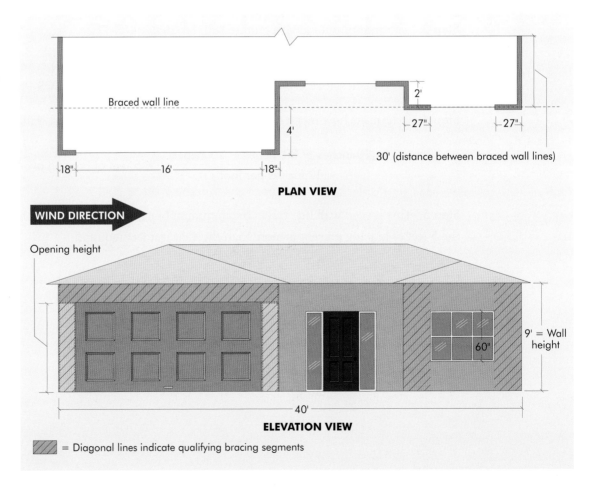

PLAN VIEW

WIND DIRECTION

ELEVATION VIEW

▨ = Diagonal lines indicate qualifying bracing segments

Solution:

Note that the two wall lines with braced wall panels are within 4 feet of the designated braced wall line, thus acceptable per R602.10.1.4.

Step 1. Determine which bracing tables, wind and/or seismic, are required for analysis.

As the wall line is a part of a single-family residence in SDC C, it is exempt from seismic design. Only wind must be considered.

Step 2. Determine how much wind bracing is required.

- From IRC Table R602.10.1.2(1) (**TABLE 6.1**), we can see that for a single-story structure, 9-foot tall wall, 90 mph Exposure B wind zone, continuously sheathed bracing, and a braced wall line spacing of 30 feet: 5 feet of bracing is required.

- Adjust values:

  ○ In accordance with Footnote c of IRC Table R602.10.1.2(1) (**TABLE 6.1**), for a single-story building with an eave-to-ridge height of 7.5 feet, by interpolation:

$$5 \text{ feet} \times 0.85 = 4.25 \text{ feet of bracing required}$$

  ○ In accordance with Footnote d of IRC Table R602.10.1.2(1) (**TABLE 6.1**), use of 9-foot tall walls permits an adjustment of 0.95:

$$4.25 \text{ feet} \times 0.95 = 4 \text{ feet of bracing required}$$

Step 3. Check garage wall minimum length requirement for Method CS-PF.

Per IRC Table R602.10.4.2, for a 9-foot tall wall, the minimum length permitted for this segment is 18 inches. (See **CHAPTER 5** for more information on Method CS-PF minimum length.)

Step 4. Check living space wall bracing minimum length requirement for Method CS-WSP.

IRC Table R602.10.4.2 permits the use of a 27-inch segment in a 9-foot tall wall adjacent to a clear opening height that is less than or equal to 64 inches. In this case, the clear opening height is the 60-inch high window in the exterior wall.

Step 5. Determine how much qualified bracing is present in the braced wall line.

$$\frac{(18 \text{ inches} + 18 \text{ inches} + 27 \text{ inches} + 27 \text{ inches})}{12 \text{ inches per foot}} = 7.5 \text{ feet of bracing available}$$

Step 6. This braced wall line meets the minimum bracing requirements for the given conditions: the 7.5 feet of bracing present is more than the 4 feet required. The distance between the centers of adjacent braced wall panels is not more than 25 feet. The existing bracing is sufficient. Note that a qualified return corner (IRC R602.10.4.4(1)) is also required at each end of each braced wall line. In lieu of a return corner, a minimum 800-lb capacity hold down must be attached to the corner studs.

Example 6.9:

Method ABW (alternate braced wall) at garage wall

Given:

- The single-family residence is in SDC B with a 90 mph Exposure B design wind speed.

- 10-foot wall height.

- Method ABW bracing with hold downs and wood structural panel sheathing is used, per IRC Section R602.10.3.2.

- Braced wall line has one story above it.

- Roof eave-to-ridge height is 5.5 feet.

- The distance between braced wall lines is 20 feet.

- *FIGURE 6.16*.

FIGURE 6.16

Example using Method ABW (alternate braced wall) at garage wall

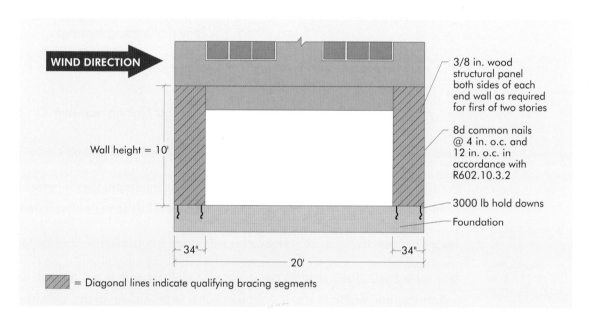

Solution:

Step 1. Determine which bracing tables, wind and/or seismic, are required for analysis.

As the wall line is a part of a single-family residence in SDC B, it is exempt from seismic design. Only wind must be considered.

Step 2. Determine how much wind bracing is required.

- From IRC Table R602.10.1.2(1) (**TABLE 6.1**), we can see that for the first of two stories, 10-foot tall wall, 90 mph Exposure B wind zone, Method WSP bracing (or any of the panel-type bracing methods, except Method GB), and a braced wall line spacing of 20 feet: <u>7.5 feet of bracing is required</u>.

- Adjust values:

  - In accordance with Footnote c of IRC Table R602.10.1.2(1) (**TABLE 6.1**), for a two-story structure with an eave-to-ridge height of 5.5 feet, by interpolation:

    7.5 feet x 0.87 = 6.53 feet of bracing required

Step 3. Check minimum braced panel geometry.

In accordance with IRC Table R602.10.3.2, for a 10-foot braced wall, a minimum panel length of 34 inches is required. Minimum braced panel geometry is met.

Step 4. Determine how much qualified bracing is present in the braced wall line.

- Note that each 34-inch segment, for the purpose of determining total required length of bracing, is equivalent to 48 inches of bracing.

$$\frac{(48 \text{ inches} + 48 \text{ inches})}{12 \text{ inches per foot}} = 8 \text{ feet of bracing available}$$

Step 5. This braced wall line meets the minimum bracing requirements for the given conditions: the 8 feet of bracing present is more than the 6.53 feet required. The existing bracing is sufficient. Note that a 3,000 lb hold down (IRC Table R602.10.3.2) is required at each end of each braced panel when using this method. Foundation reinforcement is also required in accordance with IRC Section R602.10.3.2 and IRC Figure R602.10.3.2 when Method ABW is used.

Example 6.10:

Method PFH (intermittent portal frame) with hold downs at garage wall

Given:

- The house is in SDC D$_2$ with a 110 mph Exposure C design wind speed.

- 8-foot wall height.

- Method PFH with hold downs is used, per IRC Section R602.10.3.3. Intermittent Method WSP is used on the structure.

- Braced wall line has one story above it.

- The distance between braced wall lines is 20 feet.

- Roof eave-to-ridge height is 8 feet.

- **FIGURE 6.17**.

FIGURE 6.17

Example using Method PFH (intermittent portal frame) with hold downs at garage wall

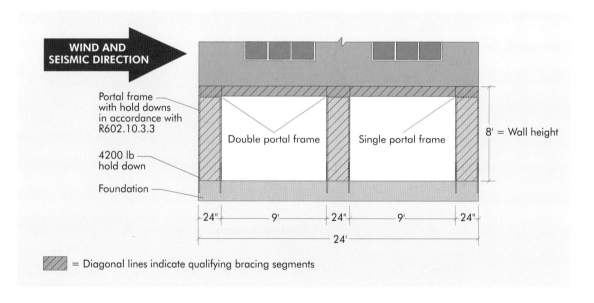

= Diagonal lines indicate qualifying bracing segments

Solution:

Step 1. Determine which bracing tables, wind and/or seismic, are required for analysis.

In accordance with IRC Section R301.2.2, this structure is located in SDC D_2 and therefore not exempt from seismic design. Both wind and seismic must be considered.

Step 2. Determine how much wind bracing is required.

- From IRC Table R602.10.1.2(1) (***TABLE 6.1***), we can see that for the first-story braced wall line in a two-story structure, 8-foot tall wall, intermittent Method WSP bracing, 110 mph wind Exposure C, and a braced wall line spacing of 20 feet: <u>11 feet of bracing is required</u>.

- Adjust values:

  - In accordance with Footnote b of IRC Table R602.10.1.2(1) (***TABLE 6.1***), a two-story Exposure C building requires an adjustment of 1.3.

    11 feet x 1.3 = 14.3 feet of wind bracing is required

  - In accordance with Footnote c of IRC Table R602.10.1.2(1) (***TABLE 6.1***), for a two-story building with an eave-to-ridge height of 8 feet, by interpolation:

    14.3 feet x 0.94 = 13.4 feet of bracing required

  - In accordance with Footnote d of IRC Table R602.10.1.2(1) (***TABLE 6.1***), use of 8-foot tall walls permits an adjustment of 0.9.

    13.4 feet x 0.90 = 12 feet of wind bracing is required

Step 3. Determine how much seismic bracing is required.

- From IRC Table R602.10.1.2(2) (***TABLE 6.2***), for a first-story wall line of a two-story structure, Method WSP bracing, SDC D_2, and a braced wall line length of 24 feet: <u>13.2 feet of bracing is required</u> by interpolation (see page 134).

- Adjust values:

  - No applicable adjustment factors.

Step 4. Determine how much qualified bracing is present in the braced wall line.

- Note that each 24-inch segment, for the purpose of determining the total length of bracing provided, is equivalent to 48 inches of bracing.

$$\frac{(48 \text{ inches} + 48 \text{ inches} + 48 \text{ inches})}{12 \text{ inches per foot}} = 12 \text{ feet of bracing available}$$

Step 5. Of the two determined bracing wall lengths, the seismic requirement is greater and controls at 13.2 feet. The braced wall line does not meet the requirement for the given conditions: the 12 feet of bracing present is less than the 13.2 feet required.

A possible solution is to reduce the length of the garage wall line to 22 feet while maintaining the same bracing.

CHAPTER 7

Braced Wall Panels and Braced Wall Lines

This chapter describes common bracing complexities, such as bracing at a distance from corners, offsets in braced wall lines, and collectors. Angled corners, a new addition for the 2009 IRC, are also introduced.

Distance between braced wall panels

Section R602.10.1.4 of the 2009 IRC requires that:

> *Braced wall panels shall be located not more than every 25 feet (7620 mm) on center and shall be permitted to begin...*

This provision, as illustrated in **FIGURE 7.1**, has been a part of the IRC since its first printing in 2000.

FIGURE 7.1

Distance between braced wall panels - minimum segment widths

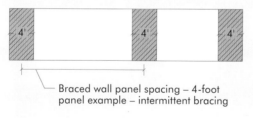

Braced wall panel spacing – 4-foot panel example – intermittent bracing

▨ Bracing Panel

The 25-foot braced wall panel spacing requirement is straightforward when 4-foot braced wall panels are used; however, the intent of the requirement should be considered when dealing with longer lengths of bracing or continuously sheathed bracing methods. The ICC Ad Hoc Wall Bracing Committee interpreted the intent to be that unbraced wall lengths longer than 21 feet (measured between adjacent bracing panel edges) are not permitted. Given this interpretation, for example, the center of a 4-foot section within an 8-foot braced wall panel can be used as the measuring point for determining the 25-foot maximum distance, as illustrated in **FIGURE 7.2**. In such a case, measuring 21 feet between bracing panels may be the easier method of determining the maximum braced wall panel spacing.

FIGURE 7.2

Distance between braced wall panels – long segment widths

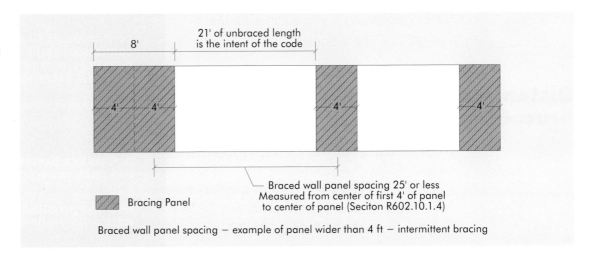

Braced wall panel spacing – example of panel wider than 4 ft – intermittent bracing

Without this interpretation, the permissible distance between bracing segments would decrease as the length of the segments increased, as illustrated in **FIGURE 7.3**. This is not the intent of the code. As long as the distance between braced wall panels does not yield unbraced wall lengths of greater than 21 feet, the IRC braced wall panel spacing requirements are satisfied.

FIGURE 7.3

Distance between braced wall panels – long segment widths

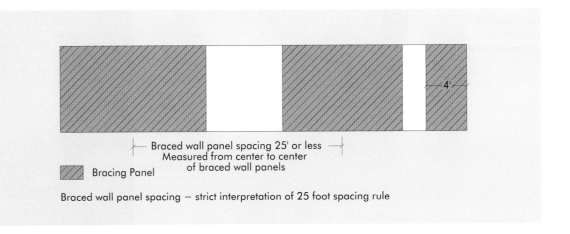

Braced wall panel spacing – strict interpretation of 25 foot spacing rule

If narrow-length braced wall panels (less than 4 feet long) are used, the 25-foot spacing is measured between the centerline of the full-height panel and the centerline of the adjacent braced wall panel(s). Methods that permit the use of narrow-length braced wall panels include:

- Method ABW (alternate braced wall)

- Method PFH (intermittent portal frame)

- Method PFG (intermittent portal frame at garage)

- Method CS-WSP (continuous wood structural panel sheathing)

- Method CS-G (wood structural panel adjacent to garage door openings and supporting roof loads only)

- Method CS-PF (continuous portal frame)

- Method CS-SFB (continuous structural fiberboard sheathing)

Note that IRC Section R602.10.3 permits traditional intermittent wall bracing methods (DWB, WSP, SFB, PCP and HPS) in lengths as short as 36 inches to be counted as bracing at a reduced effective length for bracing. When this provision is used, the 25-foot spacing measured between centerlines of the full height segments is the appropriate way to measure braced wall panel spacing.

The interpretation of a 21-foot maximum distance between bracing panel edges need not apply to narrow-length panels because it would result in a center-to-center distance that is less than the 25 feet minimum required by the IRC.

Distance between wall end and first intermittent wall panel

A change was made to this section of the code for the 2009 IRC. Section R602.10.1.4 now requires that the braced wall end distance of 12.5 feet be the combined distance for both ends of the wall. In previous editions of the IRC, the 12.5 feet end distance was permitted to be used at both ends of the wall line. For all intermittent bracing methods in Seismic Design Categories (SDC) A, B and C, braced panels may begin no more than 12.5 feet combined distance from the ends of a wall line, as shown in **FIGURE 7.4**. The 2009 IRC code reads:

> *"The total combined distance from each end of a braced wall line to the outermost braced wall panel or panels in the line shall not exceed 12.5 feet (3810mm)."*

This requirement was added to prevent the 12.5 foot end-distance provision from being used to eliminate a braced wall panel in a wall line. For example, if the 12.5 foot rule is applied at each end of a 29-foot long wall line, only a single panel is required in the wall line: one braced panel positioned 12.5 feet from each end of the wall line (12.5 feet + 4 feet + 12.5 feet = 29 feet). The new provision eliminates the possibility of such an interpretation. The original bracing provisions required bracing at each end and every 25 feet on center. A single panel in a braced wall line violates the intent of the original provision to have a minimum of two braced panels, and may not provide sufficient stability to the roof or floor diaphragm above.

FIGURE 7.4

Braced wall panel end-distance requirements

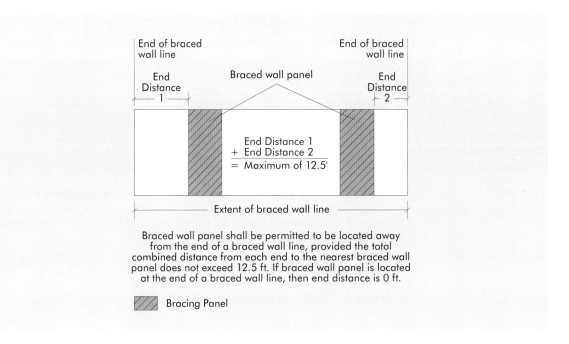

If bracing is located more than 12.5 feet from either end of a braced wall line, an engineered collector is needed to transfer the lateral loads from the roof or floor diaphragm to the braced wall panels in the braced wall line. Prescriptive information on collector selection and attachment is presented at the end of this chapter.

Distance between wall end and first intermittent wall panel in Seismic Design Categories D₀, D₁ and D₂

The corner requirements for braced wall lines changed slightly for the 2009 IRC in order to clarify the intent of the provisions. For all structures in SDC D_0, D_1 and D_2, a braced wall panel must be located at each end of all exterior braced wall lines (IRC R602.10.1.4.1).

As in the 2006 IRC, an exception is made for Method WSP bracing. Method WSP braced panels may be placed up to 8 feet in from each end of the braced wall line, provided one of the following provisions is met, as shown in *FIGURE 7.5*:

- A minimum 24-inch length wood structural panel segment attached in accordance with IRC Figure R602.10.4.4(1) is required at each side of the affected corner, or

- A minimum 1800 lb hold-down device is required on the end of each braced wall panel closest to the corner.

Note that an 8-foot maximum panel distance is permitted at each end, for a combined distance of 16 feet, if either provision is met. If two braced wall lines meet at a corner and along one wall line the first panel is displaced from the corner, then both braced wall lines lose the structural effect of the return corner. In this case, both wall lines must be anchored with an 1800 lb hold-down device.

FIGURE 7.5

For SDC D$_0$, D$_1$ and D$_2$, two options exist for bracing away from corners

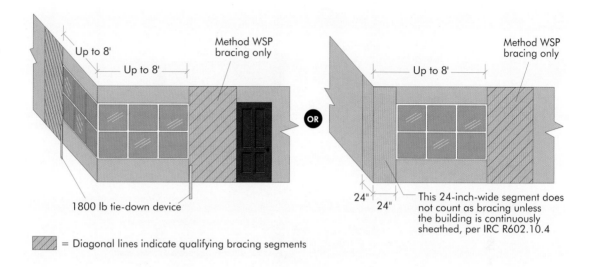

Continuously sheathed corners and end-panel distances

Continuously sheathed corners

For bracing Methods CS-WSP (IRC Section R602.10.4) and CS-SFB (IRC Section R602.10.5), a corner as shown in IRC Figure R602.10.4.4(1) (reproduced in **FIGURE 7.6**), or some other method of anchoring the wall (IRC Sections R602.10.4.4, or R602.10.5.3, respectively), is required at each end of the braced wall line. Method CS-SFB is a new addition to the IRC. Therefore, the corner requirements for this method did not appear in the corner details figure in previous versions of the IRC. Note that Method CS-SFB bracing is limited to dwellings located in SDC A, B and C and in areas with basic wind speeds below 100 mph. These limitations do not apply to Method CS-WSP. See IRC Section R602.10.5.4.

IRC Figure R602.10.4.4(1) also adds a length requirement for return corners:

- For Method CS-WSP, a 24-inch sheathed corner return is required (IRC Section R602.10.4.4), and

- For Method CS-SFB, a 32-inch sheathed corner return is required (IRC Section R602.10.5.3).

FIGURE 7.6

**Corner details
for continuous
sheathing
methods**

*Typical exterior
corner framing
for continuous
sheathing*

*IRC Figure
R602.10.4.4(1)*

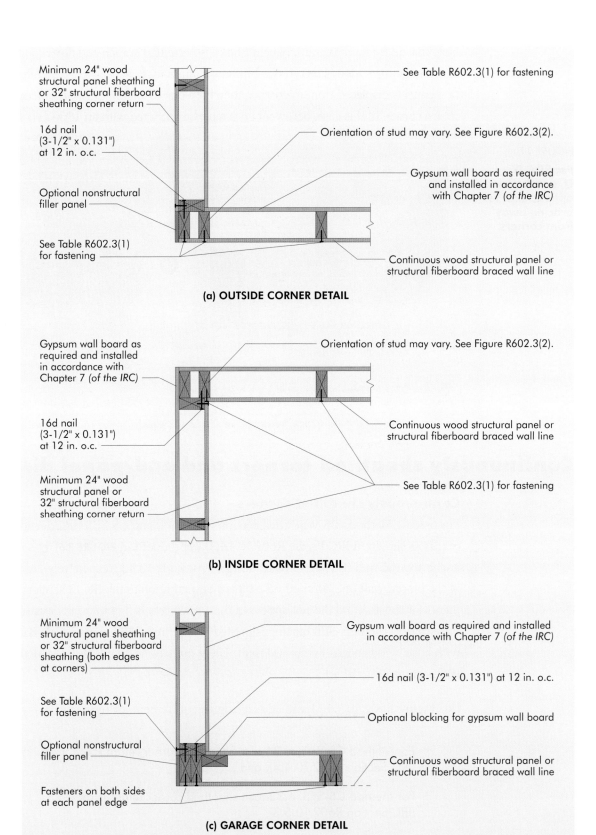

Minimum 24" wood
structural panel sheathing
or 32" structural fiberboard
sheathing corner return

16d nail
(3-1/2" x 0.131")
at 12 in. o.c.

Optional nonstructural
filler panel

See Table R602.3(1)
for fastening

See Table R602.3(1) for fastening

Orientation of stud may vary. See Figure R602.3(2).

Gypsum wall board as required
and installed in accordance
with Chapter 7 *(of the IRC)*

Continuous wood structural panel or
structural fiberboard braced wall line

(a) OUTSIDE CORNER DETAIL

Gypsum wall board as
required and installed
in accordance with
Chapter 7 *(of the IRC)*

16d nail
(3-1/2" x 0.131")
at 12 in. o.c.

Minimum 24" wood
structural panel or
32" structural fiberboard
sheathing corner return

Orientation of stud may vary. See Figure R602.3(2).

Continuous wood structural panel or
structural fiberboard braced wall line

See Table R602.3(1) for fastening

(b) INSIDE CORNER DETAIL

Minimum 24" wood
structural panel sheathing
or 32" structural fiberboard
sheathing (both edges
at corners)

See Table R602.3(1)
for fastening

Optional nonstructural
filler panel

Fasteners on both sides
at each panel edge

Gypsum wall board as required and installed
in accordance with Chapter 7 *(of the IRC)*

16d nail (3-1/2" x 0.131") at 12 in. o.c.

Optional blocking for gypsum wall board

Continuous wood structural panel or
structural fiberboard braced wall line

(c) GARAGE CORNER DETAIL

For SDC A, B and C, regardless of whether Method CS-WSP or CS-SFB is used, an 800 lb hold-down device – required to be attached between the stud at the edge of the braced wall panel closest to the corner and the foundation below – can be used in lieu of the return corner (see IRC Sections R602.10.4.4 and R602.10.5.3, and IRC Figure R602.10.4.4(3)). If two continuously sheathed braced wall lines meet at a corner, and the first braced panel is spaced away from the corner on one of the wall lines, then both braced wall lines lose the structural effect of the corner return; therefore, both braced wall lines must be anchored with an 800 lb hold-down device at the edge adjacent to the corner. The corner requirements are the same for SDC D_0, D_1 and D_2, but only Method CS-WSP may be used.

End-panel distances

The previous paragraph provided the code exceptions when corner return requirements cannot be met. Permitted bracing away from a corner in continuously sheathed wall lines is an example of when such an exception is necessary. For Method CS-WSP wall panel and corner construction, the 2009 IRC allows the first full-height braced wall segment of the continuously sheathed braced wall line to be spaced away from the end of the braced wall line, per the exception to IRC Section R602.10.4.4.

> **R602.10.4.4 Continuously sheathed braced wall panel location and corner construction.** *For all continuous sheathing methods, full-height braced wall panels complying with the length requirements of Table R602.10.4.2 shall be located at each end of a braced wall line with continuous sheathing and at least every 25 feet (7620 mm) on center. A minimum 24-inch (610 mm) wood structural panel corner return shall be provided at both ends of a braced wall line with continuous sheathing in accordance with Figures R602.10.4.4(1) and R602.10.4.4(2). In lieu of the corner return, a hold-down device with a minimum uplift design value of 800 lb (3560 N) shall be fastened to the corner stud and to the foundation or framing below in accordance with Figure R602.10.4.4(3).*
>
> **Exception:** *The first braced wall panel shall be permitted to begin 12.5 feet (3810 mm) from each end of the braced wall line in Seismic Design Categories A, B, and C and 8 feet (2438 mm) in Seismic Design Categories D_0, D_1, and D_2 provided one of the following is satisfied:*
>
> **1.** *A minimum 24 inch long (610 mm), full-height wood structural panel is provided at both sides of a corner constructed in accordance with Figure R602.10.4.4(1) at the braced wall line ends in accordance with Figure R602.10.4.4(4), or*
>
> **2.** *The braced wall panel closest to the corner shall have a hold-down device with a minimum uplift design value of 800 lb (3650 N) fastened to the stud at the edge of the braced wall panel closest to the corner and to the foundation or framing below in accordance with Figure R602.10.4.4(5).*

Note that the exception permits a 12.5-foot displacement from each end. This is different from the combined displacement of 12.5 feet permitted in IRC Section R602.10.1.4 for intermittent braced wall lines.

A similar provision applies to Method CS-SFB wall panel and corner construction (IRC Section R602.10.5.3); however, Method CS-SFB is only approved for SDC A, B and C, and the minimum full-height structural fiberboard sheathing segment at both sides of a corner is required to be 32 inches wide (instead of 24 inches, as required for Method CS-WSP).

> **R602.10.5.3 Braced wall panel location and corner construction.** *A braced wall panel shall be located at each end of a continuously sheathed braced wall line. A minimum 32-inch structural fiberboard sheathing panel corner return shall be provided at both ends of a continuously sheathed braced wall line in accordance with Figure R602.10.4.4(1) In lieu of the corner return, a hold-down device with a minimum uplift design value of 800 lb shall be fastened to the corner stud and to the foundation or framing below in accordance with Figure R602.10.4.4 (3).*
>
> **Exception:** *The first braced wall panel shall be permitted to begin 12-feet 6-inches from each end of the braced wall line in Seismic Design Categories A, B, and C provided one of the following is satisfied:*
>
> 1. *A minimum 32-inch-long, full-height structural fiberboard sheathing panel is provided at both sides of a corner constructed in accordance with Figure R602.10.4.4(1) at the braced wall line ends in accordance with Figure R602.10.4.4(4), or*
>
> 2. *The braced wall panel closest to the corner shall have a hold-down device with a minimum uplift design value of 800 lb fastened to the stud at the edge of the braced wall panel closest to the corner and to the foundation or framing below in accordance with Figure R602.10.4.4 (5).*

Note that if two continuously sheathed wall lines meet at a corner, and along one of the two wall lines, the first bracing panel is displaced from this common corner, then both braced wall lines are denied the structural effect of the corner return. In this case, both wall lines must be anchored with an 800 lb hold-down device.

Offsets in a braced wall line

Many home designs feature offsets along the braced wall line length. IRC Section R602.10.1.4 permits braced wall line offsets up to 4 feet, provided the total out-to-out offset dimension is not greater than 8 feet, as shown in **FIGURES 7.7** through **7.9**.

FIGURE 7.7

A braced wall line can have 4-foot offsets

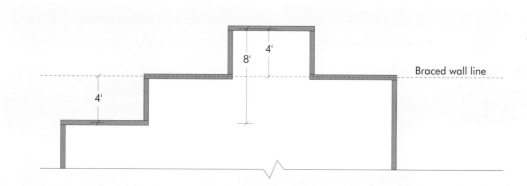

FIGURE 7.8

The code permits 8 feet total out-to-out offsets (4 feet each way) in a braced wall line

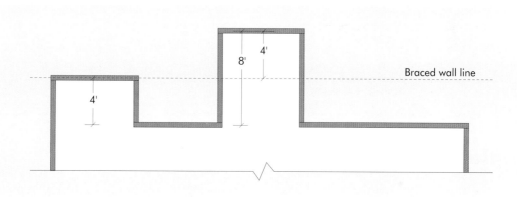

FIGURE 7.9

Offsets may occur in interior braced wall lines

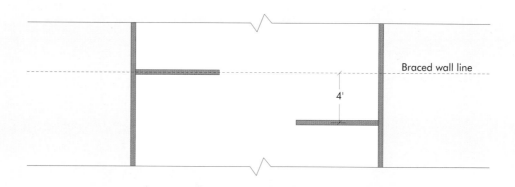

Effective (imaginary) braced wall lines

The effective (imaginary) braced wall line, a concept introduced in IRC Figure R602.10.1.4(4) of the 2009 IRC, is based on the principle that a braced wall line does not have to coincide with a physical wall line. The designated wall line location may be within a grouping of wall line sections and assumed to provide a line of lateral resistance somewhere near the center of the wall sections. This concept is covered in depth in **CHAPTER 3** and illustrated in **FIGURE 7.10**. The offset rules discussed previously apply to the effective (imaginary) braced wall line location.

FIGURE 7.10

Effective (imaginary) braced wall line

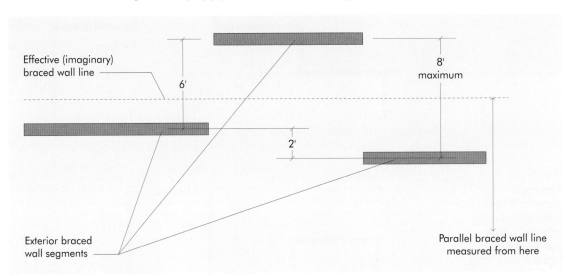

Angled corners

New for the 2009 IRC is a section on angled corners (R602.10.1.3), reproduced below and in **FIGURE 7.11**. This provision allows the sheathing in a diagonal wall to be "counted" towards the total bracing length for a single braced wall line. Note that the angled portion of the wall line must be less than or equal to 8 feet in length. If it is greater than 8 feet in length, it must be considered its own braced wall line.

> **R602.10.1.3 Angled corners.** *At corners, braced wall lines shall be permitted to angle out of plane up to 45 degrees with a maximum diagonal length of 8 feet (2438 mm). When determining the length of bracing required, the length of each braced wall line shall be determined as shown in Figure R602.10.1.3. The placement of bracing for the braced wall lines shall begin at the point where the braced wall line, which contains the angled wall adjoins the adjacent braced wall line (Point A as shown in Figure R602.10.1.3). Where an angled corner is constructed at an angle equal to 45 degrees and the diagonal length is no more than 8 feet (2438 mm) in length, the angled wall may be considered as part of either of the adjoining braced wall lines, but not both. Where the diagonal length is greater than 8 feet (2438 mm), it shall be considered its own braced wall line and be braced in accordance with Section R602.10.1 and methods in Section R602.10.2.*

FIGURE 7.11

Angled corners

*IRC Figure
R602.10.1.3*

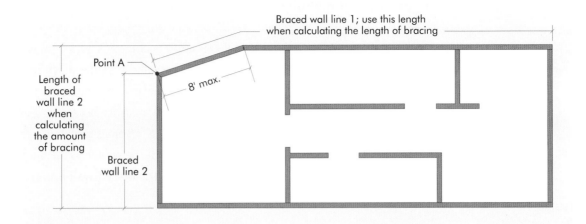

The angled corner provision was adopted simultaneously with the division of the previously joint wind and seismic bracing table into two separate bracing tables. Unfortunately, the annotations in **FIGURE 7.11** still reflect the methodology used in previous editions of the IRC.

Note that the provision does not permit "double dipping": you cannot count the bracing on an angled corner towards the total bracing length required for both braced wall lines connected to the corner.

What if the angled corner contains the entryway of the house and there is not sufficient space for bracing panels?

The length of the angled entryway still counts towards the length of the wall line, and the 12.5 foot rule (IRC Section R602.10.1.4) also applies. As the length of the angled corner shown in **FIGURE 7.12** is less than 12.5 feet (less than or equal to 8 feet in SDC D_0, D_1 and D_2, per IRC Section R602.10.1.4.1), bracing panels are not required in the angled section, provided that the other requirements of IRC Section R602.10.1.4 are met.

FIGURE 7.12

Insufficient room in angled portion of wall to permit bracing

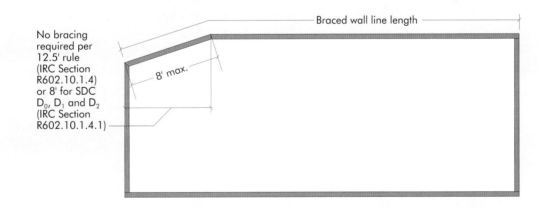

What happens if the angled wall segment is greater than 12.5 feet? Whenever the angled section of the wall is greater than 8 feet, the angled corner provision of IRC Section R602.1.1.3 no longer applies and the angled section is considered its own wall line that must be braced accordingly, as shown in **FIGURE 7.13**. This can be a difficult proposition when using the 2009 IRC because of the need for "braced wall line spacing" per both the wind and seismic bracing tables. If no parallel angled line is available to measure between, then determining the correct distance to use for the braced wall line spacing becomes a matter of interpretation. A reasonable solution is to run a perpendicular line from the center of the angled wall until it intersects a braced wall line and use that distance as the braced wall line spacing.

FIGURE 7.13

Angled portion of wall greater than 8 feet

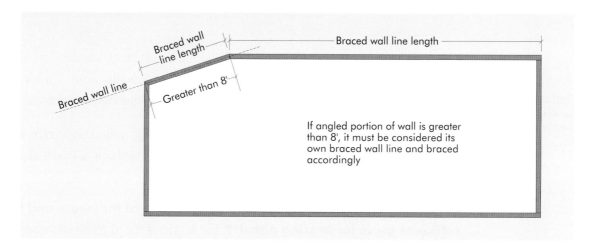

If there are no bracing elements in the angled segment and the first braced wall panel is a distance greater than 12.5 feet from Point A (as shown in **FIGURE 7.11**), then a collector may be used as an engineered solution (see following section on collectors). In this case, however, since a continuous top plate is not available to act as a collector due to the change in direction of the wall, metal straps will be required to make the tension splice.

Collectors

When bracing is placed farther from the end of the braced wall line than specified (8 feet in SDC D_0, D_1 and D_2 – Methods WSP and CS-WSP only – or 12.5 feet for all bracing methods in SDC A, B and C) the limits of the IRC are exceeded. In the 2009 IRC, references to collector design (previously included in R602.10.11.3 in the 2006 IRC) were omitted because they were considered to be redundant with Section R104.11, which permits approved designed solutions for any situation that falls outside of the scope of the IRC. A collector (sometimes called a "drag strut") is an engineered element used for such applications. Prescriptive information is presented on the following pages. (Engineering information is provided in the free APA publication, *Technical Topics: Collector Design for Bracing in Conventional Construction*, Form TT-102. Visit www.apawood.org.)

Note that collector design, like any alternate proposed in accordance with IRC Section R104.11, must be approved by the building official. It might prove prudent to contact your local building official during the drawing-development phase of the project.

What is a collector and what does it do?

The collector is a part of the lateral load path. Just as a beam in the vertical load path accumulates (or "collects") gravity loads and carries them over an opening to the jack studs on each side, the collector has a similar function in the lateral load path. It collects the lateral loads from the roof or floor diaphragm (roof sheathing and floor sheathing, respectively) above an opening and distributes them into the braced wall panels on either side (or one side, as circumstances dictate) of this opening in the braced wall line below. Thus, a collector is needed to evenly distribute the load among the bracing panels in a given wall line.

What does a collector look like and how do I design one?

In the type of structures considered by the IRC, the collector is normally already in place in the form of the wall double top plate or top plate/header combination (see **FIGURE 7.14**).

FIGURE 7.14

First bracing panel shown 14 feet from corner

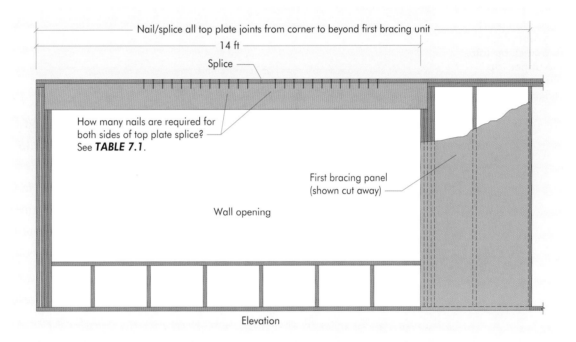

Nail/splice all top plate joints from corner to beyond first bracing unit

14 ft

Splice

How many nails are required for both sides of top plate splice? See **TABLE 7.1**.

First bracing panel (shown cut away)

Wall opening

Elevation

In terms of load path, the roof or upper-floor framing is attached to the wall top plate. The top plate is also a part of the framing for the bracing panels immediately below. This connection between the roof or floor framing and the wall ensures that the horizontal lateral loads (from the floor or roof sheathing/diaphragm above) are distributed into the bracing panels (vertical lateral load resisting elements). The IRC Table R602.3(1), Item 13 attachment requirement for double top plate splices (minimum 24-inch offset with eight 16d nails in splice area) ensures continuity along the 8- or 12.5-foot corner distance limit. Beyond these limits, the prescriptive top plate splice cannot be counted on to transfer greater loads. The IRC-required fastener schedule (IRC Table R602.3(1), Item 15) for the wall bottom plate of the bracing panel (three 16d face nails into joist or blocking every 16 inches on center) completes the load path to the foundation or floor below.

Choosing a collector

The good historical performance of traditional framing methods in conventional construction has led to code provisions that permit the first bracing panel to be placed away from the corner up to a set distance without requiring the builder to be concerned with the collector design. For distances in excess of those permitted by the code, the material properties of a single 2x4 top plate are generally adequate to distribute the applied load for conventional construction. The difficulty lies in providing sufficient attachment between the upper and lower top plates (or the upper top plate and header below) to transfer the load at "splice locations" where joints in the double top plate occur (see **FIGURE 7.15**).

FIGURE 7.15

Splice at top panel required to transfer load across joint in upper top plate to lower plate and back up into the upper top plate after the splice

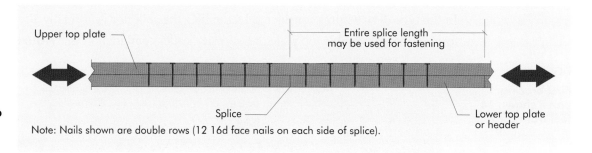

The IRC requires that a minimum splice nailing of eight 16d face nails (IRC Table R602.3(1), Item 13) – or in the case of SDC D_0, D_1 and D_2 and illustrated in **FIGURE 7.15**, twelve 16d face nails (IRC Section R602.10.1.5, Exception) – be placed on both sides of the top plate joint. Note that each side of the joint is a "lapped area". The normal double top plate face nailing requirements (IRC Table R602.3(1), Item 12) may be counted toward the minimum number, but the minimum number must be met. Note that IRC Section R602.10.6.1 is more specific in its treatment of top plate lap splices, requiring that:

> *"...at least eight 16d nails on each side of the splice."*

What is the length of the top plate that must be spliced?

Since the purpose of the collector is to "collect" the load from the structure above and transfer it into the bracing panel a distance away from the corner, the spliced top plate must run from the building corner to at least the far side of the first bracing panel.

The top plate splice design parameters can be represented in tabular form, as shown in **TABLE 7.1**.

TABLE 7.1

Top plate splice design table[a]

L (Distance between corner and beginning of first bracing panel in ft)		8 ft[c]	10 ft[c]	12 ft[c]	12.5 ft[c]	14 ft	16 ft	18 ft	20 ft
N (Number of 16d box nails required on each side of top plate butt joint)	Basic Wind Speed < 110 mph SDCs A - C	8 nails (per IRC Table R602.3.(1), Item 13)				21 nails	24 nails	27 nails	30 nails
	SDCs D₀, D₁, D₂	8 nails	15 nails	18 nails	19 nails				
	SDCs D₀, D₁, D₂ and IRC Table R602.10.1.5, Exception, is used to increase braced wall line spacing to 35 feet.	12 nails	15 nails	18 nails	19 nails				
Minimum double top plate size, species, and grade[b]	2x4	SPF Stud Grade						DF Stud Grade or SPF #2	DF #2 or SPF #2
	2x6	SPF Stud Grade							

a. If 16d common nails (0.162 x 3-1/2 in.) are used, the number of nails in the above table may be multiplied by 0.73.

b. Once a size, species and grade is selected from the table, other sizes, species and grades listed to the right of the selected grade, or of greater strength per the NDS, may be substituted.

c. Engineered collector not required under the 2009 IRC for wind or SDCs A–C when bracing begins within 12-1/2 ft of end of wall or in SDCs D₀-D₂ when bracing starts within 8 ft of end of wall.

Design example: A home designer wants to place a bracing panel 14 feet away from a corner. The wall framing is 2x4. Select the lumber species and grade, and detail the collector. See **FIGURE 7.14**.

Solution: **TABLE 7.1** shows that for an L (distance between corner and beginning of first bracing panel in feet) of 14 feet, the top plate splice made of 2x4 SPF (spruce-pine-fir) stud grade requires 21 16d box nails on each side of the top plate splice, as illustrated in **FIGURE 7.16**. If common nails are used, 16 16d common nails are required on each side of the top plate splice, per Footnote a of **TABLE 7.1** (21 nails x 0.73 = 16 nails).

Where the splice occurs over a lower top plate, a minimum 4-foot overlap would, for this example, provide enough room for the required fasteners without causing splitting of the plates. (Note that although a 4-foot splice overlap would theoretically provide enough room, staggering each row of nails, blunting the tips of the nails, and/or using three rows of fasteners is recommended if splitting appears to be a problem given the wood resource used.) Where the splice occurs over a header, a lower top plate is not necessary. Note that using longer plates (12-16 feet) minimizes the number of plate splices required.

FIGURE 7.16

Top plate splice detail for example

21 16d box nails shown each side of joint

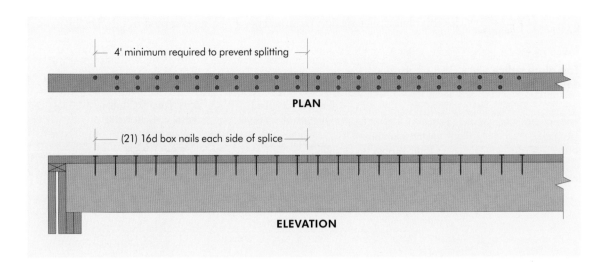

4' minimum required to prevent splitting

PLAN

(21) 16d box nails each side of splice

ELEVATION

CHAPTER 8

Bracing Connection and Foundation Anchorage Details

This chapter describes the IRC connection requirements between braced wall panels and the portions of the structure both above and below. Without the connections required to complete the lateral load path, the purpose of wall bracing will not be realized and the structure will not be able to resist lateral loads.

The braced wall connection requirements in the 2009 IRC include added load path details with illustrations for both raised- and low-heel roof trusses. Foundation anchorage information has also been added, as well as essential details for providing reinforcement to narrow-width braced panels for garage walls when masonry block foundations are used.

In addition to the requirements of the IRC, related information and suggestions regarding the proper application of bracing are discussed in this chapter.

Bracing connection details

Braced wall panel joints

2009 IRC Section R602.10.8 requires that all vertical and horizontal joints in panel sheathing used for bracing occur over and be attached to common framing. This is to ensure that the bracing performs as intended when subjected to lateral loads. The requirement for horizontal joints to be blocked when they occur between the top and bottom plates has some exceptions:

- The first exception states that the horizontal joints of sheathing panels not used as braced panels do not have to be blocked. However, panel manufacturers or panel associations may recommend blocking the horizontal joints of sheathing panels regardless of whether they are used as braced panels. For example, brittle finishes (such as stucco) over panel wall sheathing may require blocked horizontal panel edges to prevent cracking, even when the panels are not used for wall bracing.

- When horizontal joints of braced wall panels are not blocked, the effectiveness of the bracing is reduced and additional bracing is required. Consequently, for Methods WSP, SFB, PBS and HPS, horizontal blocking of the braced wall panels may be eliminated, provided that twice the minimum required bracing length is used in the wall line. The purpose of this exception is to provide a construction alternative to installing horizontal blocking when circumstances permit.

- The required bracing length for Method GB (gypsum board) is based on its vertical application with horizontal joints occurring over framing or blocking. Method GB bracing is considerably stronger when installed horizontally. The third exception permits Method GB bracing to be unblocked when it is installed horizontally.

Connection of braced wall panels to floor/ceiling framing

IRC Table R602.3(1) provides the basic nailing schedule for attaching braced wall panels to the floor below and the ceiling or floor above. As an example, the sole plate (or bottom plate) of the braced wall panels is required to be nailed with three 16d nails every 16 inches on center into the joist or blocking below the braced wall panels. See **TABLE 8.1**.

Note that the nails do not have to be clustered every 16 inches on center: there just has to be three nails in every 16 inch-length of braced wall panel. In fact, clustering the nails may not be a good idea if, for example, they are going into a rim board. In other cases, clustering may be needed, such as an interior braced wall parallel to and situated between floor framing: clustering is necessary to attach the braced wall to the blocking. See **FIGURE 8.2**.

TABLE 8.1

Attachment of braced wall panels

IRC Table R602.3(1) (Excerpt) Fastener schedule for structural members

Item	Description Of Building Elements	Member And Type Of Fastener	Spacing Of Fasteners
15	Sole plate to joist or blocking, at braced wall panels	3-16d (3-1/2" x 0.135")	16" o.c.

For a braced wall panel to provide lateral load resistance to the structure, it is essential that lateral loads transfer into the panel at its top and out to the framing members below at its bottom. Correct connections are necessary to complete this lateral load path and are addressed in the following new braced wall panel connection provisions. New figures are also provided in the 2009 IRC to support these provisions (see **FIGURES 8.1** and **8.2**).

> **R602.10.6 Braced wall panel connections.** *Braced wall panels shall be connected to floor framing or foundations as follows:*
>
> **1.** *Where joists are* **perpendicular** *to a braced wall panel above or below, a rim joist, band joist or blocking shall be provided along the entire length of the braced wall panel in accordance with Figure R602.10.6(1) (**FIGURE 8.1**). Fastening of top and bottom wall plates to framing, rim joist, band joist and/or blocking shall be in accordance with Table R602.3(1).*

FIGURE 8.1

Connections for braced wall panels perpendicular to floor/ceiling framing

IRC Figure R602.10.6(1) Braced wall panel connection when perpendicular to floor/ceiling framing

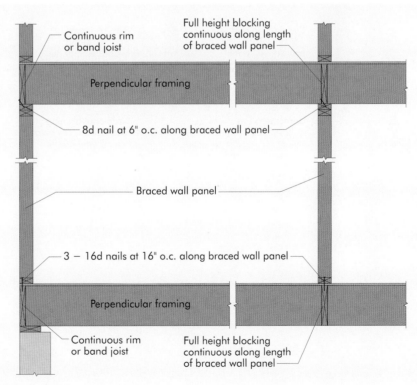

For SI: 1 inch = 25.4 mm

2. *Where joists are **parallel** to a braced wall panel above or below, a rim joist, end joist or other parallel framing member shall be provided directly above and below the braced wall panel in accordance with Figure R602.10.6(2) (**FIGURE 8.2**). Where a parallel framing member cannot be located directly above and below the panel, full-depth blocking at 16 inch (406 mm) spacing shall be provided between the parallel framing members to each side of the braced wall panel in accordance with Figure R602.10.6(2). Fastening of blocking and wall plates shall be in accordance with Table R602.3(1) and Figure R602.10.6(2).*

FIGURE 8.2

Connections for braced wall panels parallel to floor/ceiling framing

IRC Figure R602.10.6(2) Braced wall panel connection when parallel to floor/ceiling framing

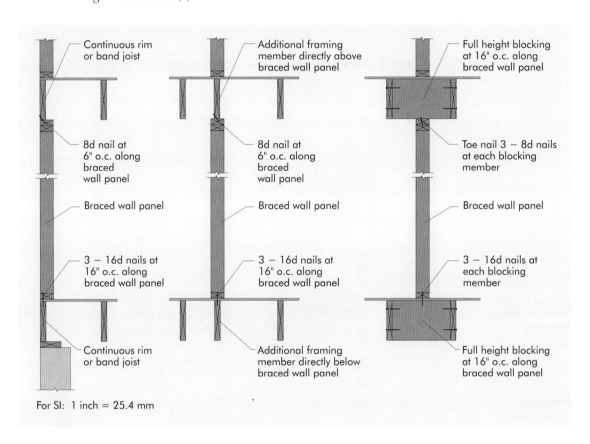

For SI: 1 inch = 25.4 mm

3. *Connections of braced wall panels to concrete or masonry shall be in accordance with Section R403.1.6.*

Connection of braced wall panels to floor/ceiling framing for Seismic Design Categories D_0, D_1 and D_2

Structures in higher seismic zones need to have the capacity to withstand greater loads; therefore, additional provisions are required to ensure that these loads are transferred through the braced wall panels to the foundation. The requirements of IRC Section R602.10.6.1 ensure the top plate lap splices and foundation connections have the needed higher load capacity.

The requirement for eight 16d nails on each side of the lap splice in SDCs D_0, D_1 and D_2 is reiterated in IRC Section R602.10.6.1. Note that in the 2009 IRC, Table R602.3(1), Item 13, requires a 24-inch overlap to provide sufficient length to put in the eight 16d nails. See **FIGURE 8.3**.

R602.10.6.1 Braced wall panel connections for Seismic Design Categories D_0, D_1 and D_2. Braced wall panels shall be fastened to required foundations in accordance with Section R602.11.1, and top plate lap splices shall be face-nailed with at least eight 16d nails on each side of the splice.

The collector design information in **CHAPTER 7** is appropriate for collector splices when the first bracing panel is displaced from the corner by more than 8 feet.

FIGURE 8.3

Top plate splice for SDC D_0, D_1 and D_2

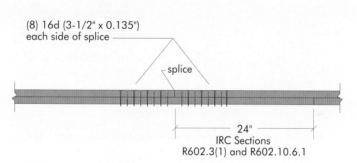

(8) 16d (3-1/2" x 0.135") each side of splice

splice

24"
IRC Sections
R602.3(1) and R602.10.6.1

Top plate splice detail for interior braced wall line

Connection of braced wall segments to roof framing – all applications

The 2009 IRC includes new and important provisions for the connection between braced wall panels and roof framing. Although previous versions of the IRC did not address this connection, complex roof shapes used in modern design have necessitated prescriptive connection details to ensure an effective load path exists.

These connection details represent a simple principle: braced wall lines extend from diaphragm to diaphragm and must be connected at both top and bottom. The roof and floor sheathing are the structural diaphragms of the building. IRC Figures R602.10.6.2(1)-(3) illustrate the required connections and blocking between braced wall panels and roof framing, which are summarized in **TABLE 8.2**.

TABLE 8.2

Connection and blocking requirements between braced wall panels and roof framing

Seismic Design Category & Wind Speed	Distance (bottom of roof sheathing to top of top plate) (See *FIGURE 8.4*)	Blockingª
SDC A, B, C and wind speed less than 100 mph	9.25" or less	Not Required per IRC Section R602.10.6.2, Item 2. Roof framing attached per IRC Section R602.3(1).
	Greater than 9.25" to 15.25"	Required per IRC Section R602.10.6.2, Item 2 and Figure R602.10.6.2(1)
SDC D₀, D₁, D₂ or wind speed 100 mph or greater	15.25" or less	Required per IRC Section R602.10.6.2, Item 3 and Figure R602.10.6.2(1)
All SDCs and wind speeds	15.25" to 48"	Required per IRC Section R602.10.6.2, Item 4 or Figures R602.10.6.2(2) or R602.10.6.2(3)

a. Rafter or truss connection to top plate per IRC Table R602.3(1).

FIGURE 8.4

Distance from top plate to bottom of roof sheathing for providing connection requirements and blocking per *TABLE 8.2*.

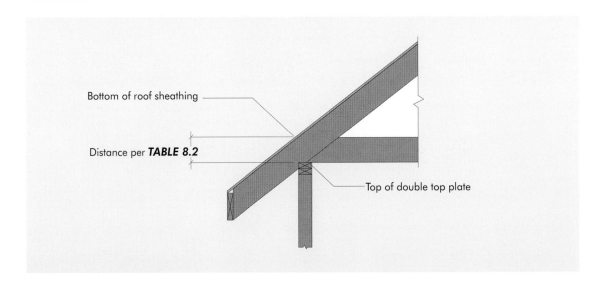

Bottom of roof sheathing

Distance per **TABLE 8.2**

Top of double top plate

IRC Section R602.10.6.2, Item 1, requires parallel roof framing to be attached to the top of braced wall panels in accordance with IRC Table R602.3(1) for all wind speeds and SDCs covered by the IRC. This is done to ensure a proper load path from the roof diaphragm into the braced wall panels below. Because this section deals with the bracing provisions alone, the requirements only specify attachment at braced wall panels. Note that the required attachment specified in IRC Table R602.3(1) is appropriate for all locations in the structure. Ultimately, all roof framing must be attached to the supporting walls in accordance with the minimum provisions of IRC Table R602.3(1).

IRC Section R602.10.6.2, Item 2, provides what is essentially an exemption from required blocking for SDC A, B and C and wind speeds less than 100 mph, where the distance from the top of the rafters or roof trusses to the perpendicular top plates is 9.25 inches or less, provided the rafters or roof trusses are connected to the top plates of braced wall lines in accordance with IRC Table R602.3(1).

IRC Figure R602.10.6.2(1) (shown in **FIGURE 8.5**) illustrates the required detailing and connection for relatively low-heel trusses and rafters (less than or equal to 15.25 inches between the bottom of roof diaphragm sheathing and the top of double top plate). For such applications, the addition of solid lumber blocking is sufficient to prevent the trusses or conventional framing members from rolling over when subjected to wind and/or seismic loads. Attachment per IRC Table R602.3(1), Items 1, 2 and 5 – nailing at the bottom of the roof framing and blocking – is required to transfer the lateral forces (that push the roof framing along the top plate) from the roof sheathing (or diaphragm) into the braced wall line. The 2-inch gap at the top of the blocking permits required roof ventilation.

PARTIAL-HEIGHT BLOCKING

The continuous 2-inch gap at the top of the blocking or blocking panel shown in IRC Figures R602.10.6.2(1) and R602.10.6.2(2) is new to the 2009 IRC. In the past, lateral blocking required to resist racking or cross-grain bending of the roof framing was specified as full-height from top of top plate to bottom of sheathing. Where ventilation openings were required, an engineered solution was necessary.

A 2002 HUD-PATH investigation of conventional roof framing connections showed that lateral blocking for typical rafter/ceiling-joist assemblies and low-heel trusses was not required in low-wind, low-seismic conditions. This testing forms the basis for the requirements of IRC Section R602.10.6.2, Item 2.

For deeper assemblies and higher-load conditions, it is possible to engineer "partial-height" blocking that does not extend the full depth of the rafter or truss heel. Guidance for estimating cross-grain bending stresses in natural lumber is provided in the USDA FPL Wood Handbook. Using a conservative estimate of the cross-grain bending stress, it was determined that a continuous 2-inch gap at the top of solid blocking or a blocking panel would not compromise the structural performance of the roof framing for loads within the scope of the IRC.

IRC Figure R602.10.6.2(2) (shown in *FIGURE 8.6*) illustrates an alternate connection between a braced wall panel and roof sheathing (or diaphragm) that, while less direct, will provide adequate transfer of lateral forces from the roof diaphragm to the braced wall line. This detail does not provide lateral stability of the trusses at the bearing point (see truss installation diagrams) or the required roof venting. While the code permits various bracing methods for this application (IRC Figure R602.10.6(2), Note a), the user must be careful to specify only those methods that have sufficient weather resistance or provide a covering with an approved exterior finish. Note that the edge nailing requirement for the bracing method used at the soffit is provided in IRC Table R602.10.2 and is not necessarily the same as the roof sheathing edge nailing, unless the same material is used for both applications.

IRC Figure R602.10.6.2(3) (shown in *FIGURE 8.7*) illustrates a connection applicable for raised-heel trusses of up to 4 feet in height. In addition to transferring the lateral load from the roof sheathing (or diaphragm) to the braced wall line, this method does address lateral stability of the trusses at the bearing point. Ventilation is provided by a gap of up to 2 inches at the top of braced panels between trusses. While the code permits various bracing methods for this application (IRC Figure R602.10.6(3), Note a), the user must be careful to specify only those methods that have sufficient weather resistance or provide a covering with an approved exterior finish. Closed soffit construction may also be used to protect bracing materials at this location. Note that the edge nailing requirement for the specific bracing method used to construct this option is provided in IRC Table R602.10.2 and is not necessarily the same as the roof sheathing edge nailing, unless the same material is used for both applications.

This portion of the 2009 IRC is reproduced below:

R602.10.6.2 Connections to roof framing. *Exterior braced wall panels shall be connected to roof framing as follows.*

1. *Parallel rafters or roof trusses shall be attached to the top plates of braced wall panels in accordance with Table R602.3(1).*

2. *For SDC A, B and C and wind speeds less than 100 mph (45 m/s), where the distance from the top of the rafters or roof trusses and perpendicular top plates is 9.25 inches (235 mm) or less, the rafters or roof trusses shall be connected to the top plates of braced wall lines in accordance with Table R602.3(1) and blocking need not be installed. Where the distance from the top of the rafters and perpendicular top plates is between 9.25 inches (235 mm) and 15.25 inches (387 mm) the rafters shall be connected to the top plates of braced wall panels with blocking in accordance with Figure R602.10.6.2(1) (**FIGURE 8.5**), and attached in accordance with Table R602.3(1). Where the distance from the top of the roof trusses and perpendicular top plates is between 9.25 inches (235 mm) and 15.25 inches (387 mm) the roof trusses shall be connected to the top plates of braced wall panels with blocking in accordance with Table R602.3(1).*

3. *For SDC D_0, D_1 and D_2 or wind speeds of 100 mph (45 m/s) or greater, where the distance between the top of rafters or roof trusses and perpendicular top plates is 15.25 inches (387 mm) or less, rafters or roof trusses shall be connected to the top plates of braced wall panels with blocking in accordance with Figure R602.10.6.2(1) (**FIGURE 8.5**), and attached in accordance with Table R602.3(1).*

FIGURE 8.5

Braced wall panel connection – low-heel trusses

IRC Figure R602.10.6.2(1) Braced wall panel connection to perpendicular rafters

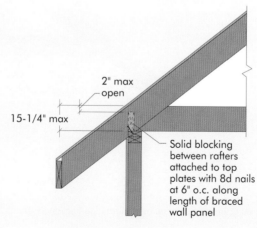

2" max open

15-1/4" max

Solid blocking between rafters attached to top plates with 8d nails at 6" o.c. along length of braced wall panel

For SI: 1 inch = 25.4 mm

4. For all Seismic Design Categories and wind speeds, where the distance between the top of rafters or roof trusses and perpendicular top plates exceeds 15.25 inches (387 mm), perpendicular rafters or roof trusses shall be connected to the top plates of braced wall panels in accordance with one of the following methods:

   1. In accordance with Figure R602.10.6.2(2) (**FIGURE 8.6**),

   2. In accordance with Figure R602.10.6.2(3) (**FIGURE 8.7**),

   3. With full height engineered blocking panels designed for values listed in American Forest and Paper Association (AF&PA) Wood Frame Construction Manual for One- and Two-Family Dwellings (WFCM). Both the roof and floor sheathing shall be attached to the blocking panels in accordance with Table R602.3(1).

   4. Designed in accordance with accepted engineering methods.

Lateral support for the rafters and ceiling joists shall be provided in accordance with Section R802.8. Lateral support for trusses shall be provided in accordance with Section R802.10.3. Ventilation shall be provided in accordance with R806.1.

FIGURE 8.6

Braced wall panel connection – boxed soffit method

IRC Figure R602.10.6.2(2) Braced wall panel connection option to perpendicular rafters or roof trusses

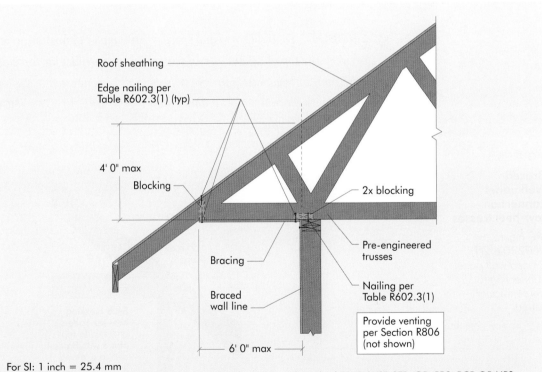

For SI: 1 inch = 25.4 mm
a. Methods of bracing shall be as described in Section R602.10.2 method DWB, WSP, SFB, GB, PBS, PCP OR HPS

FIGURE 8.7

**Braced
wall panel
connection –
maximum height
4 feet**

*IRC Figure
R602.10.6.2(3)
Braced wall panel
connection option
to perpendicular
rafters or roof
trusses*

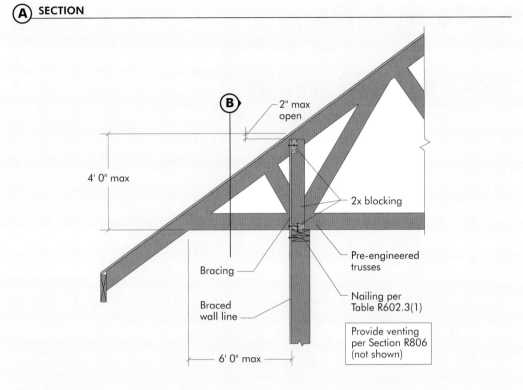

Ⓐ **SECTION**

Ⓑ

2" max
open

4' 0" max

2x blocking

Bracing

Pre-engineered
trusses

Braced
wall line

Nailing per
Table R602.3(1)

Provide venting
per Section R806
(not shown)

6' 0" max

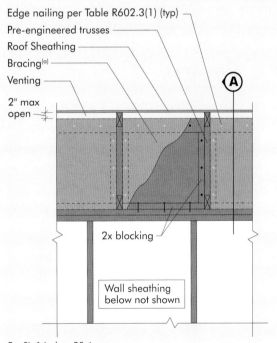

Ⓑ **ELEVATION**

Edge nailing per Table R602.3(1) (typ)
Pre-engineered trusses
Roof Sheathing
Bracing[a]
Venting

2" max
open

Ⓐ

2x blocking

Wall sheathing
below not shown

For SI: 1 inch = 25.4 mm
(a) Methods of bracing shall be as described in Section
R602.10.2 Methods DWB, WSP, SFB, GB, PBS, PCP or HPS

Bracing support, anchorage and foundations

Braced wall panel support

Section R602.10.7 of the 2009 IRC contains requirements for several braced wall line special circumstances not addressed in previous editions of the IRC. Included are provisions for cantilevered floor joists, elevated post and pier foundations, and masonry stem walls.

Floor cantilevers

IRC Section R502.3.3 provides the basic cantilever requirements, and Section R602.10.7 further clarifies those requirements for use with braced walls.

IRC Section 502.3.3 restricts floor cantilevers for general use to the depth of the floor joist. Thus a 2x10 floor joist can cantilever out 9-1/4 inches for all cases covered by the IRC without other considerations. IRC Table R502.3.3(1) permits greater length cantilevers if the structure above the cantilever is limited to a single light-frame wall and roof. The length of these cantilevers can be as long as 48 inches under certain circumstances.

IRC Section R602.10.7, Item 1 refines these provisions for use in conjunction with braced wall lines placed over cantilevers. Cantilevers supporting braced wall lines must have solid blocking at the nearest bearing wall locations. In SDCs A, B and C, if the cantilever is limited to 24 inches or less, a full-height rim joist at the end of the cantilevered joists may be used instead of solid blocking. See **FIGURE 8.8**.

While IRC Section R602.10.7, Item 1 provides no guidance on cantilever detailing for wind speeds or moderate-to-high seismic loads, the requirements in IRC Section R502.3.3 and Table R503.3.3(1) must still be followed. As a result, there is a conflict between R602.10.7, Item 1 and Table R502.3.3(1). Footnote g of IRC Table R502.3.3(1) requires BOTH the blocking at the support AND the continuous full-height rim board at the end of the cantilever, regardless of the basic wind speed or SDC. The exemption in IRC Section R602.10.7 for SDCs A, B and C seems incomplete as no similar exemption for areas of low wind is provided. This exemption also begs the question, "What if the basic wind speed is 105 mph in a low seismic SDC? Is it still permissible to delete the blocking?" While the final decision is ultimately made by the local building official, it is conservative to assume that the requirements of IRC Section R502.3.3 would control.

FIGURE 8.8

Braced wall lines over cantilever foors

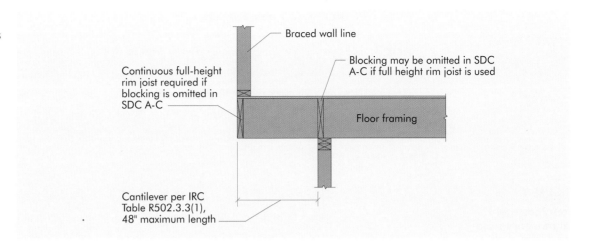

Braced wall line

Blocking may be omitted in SDC A-C if full height rim joist is used

Continuous full-height rim joist required if blocking is omitted in SDC A-C

Floor framing

Cantilever per IRC Table R502.3.3(1), 48" maximum length

Elevated post and pier foundations

Elevated post and pier foundations generally provide little lateral support to the structure above. Lateral support instead typically comes from diagonal bracing or other elements of the foundation; however, there are ways of cantilevering the posts of the foundation by burying the ends deeply into the ground, so that they develop lateral load resisting characteristics.

It is difficult to accurately and fully present all of the post and pier lateral support methods in one set of prescriptive provisions. For this reason, post and pier foundations supporting braced wall panels must be engineered in accordance with the IBC or referenced documents, as stated in the IRC R602.10.7, Item 2:

> *Elevated post or pier foundations supporting braced wall panels shall be designed in accordance with accepted engineering practice.*

Masonry stem walls

Past field problems have made it clear that free-standing unreinforced masonry foundations adjacent to garage doors (or similar openings) may not perform well with narrow bracing options. So for the 2009 IRC, the ICC Ad Hoc Committee on Wall Bracing and the National Concrete Masonry Association developed IRC Figure R602.10.7, shown in **FIGURE 8.9**, along with provisions for the reinforcement of masonry stem walls that support braced wall panels. The provisions are intended for masonry stem walls adjacent to garage doors or similar openings.

- The reinforcement details in IRC Figure R602.10.7 are appropriate for masonry stem walls that are up to 4 feet in length and not more than 4 feet in height.

- If the masonry stem walls are taller than 4 feet, an engineered design of the reinforcement is required.

- If the masonry stem walls are longer than 4 feet, this specific reinforcement is not necessary (standard construction in accordance with IRC Section R403.1 is sufficient).

- Masonry stem walls are not permitted to support Method ABW and PFH bracing unless specifically engineered for such applications.

R602.10.7 Braced wall panel support.

Item 3. *Masonry stem walls with a length of 48 inches (1220 mm) or less supporting braced wall panels shall be reinforced in accordance with Figure R602.10.7. Masonry stem walls with a length greater than 48 inches (1220 mm) supporting braced wall panels shall be constructed in accordance with Section R403.1. Braced wall panels constructed in accordance with Sections R602.10.3.2 and R602.10.3.3 shall not be permitted to attach to masonry stem walls.*

FIGURE 8.9

Reinforcement of masonry stem walls supporting bracing elements

IRC Figure R602.10.7 Masonry stem walls supporting braced wall panels

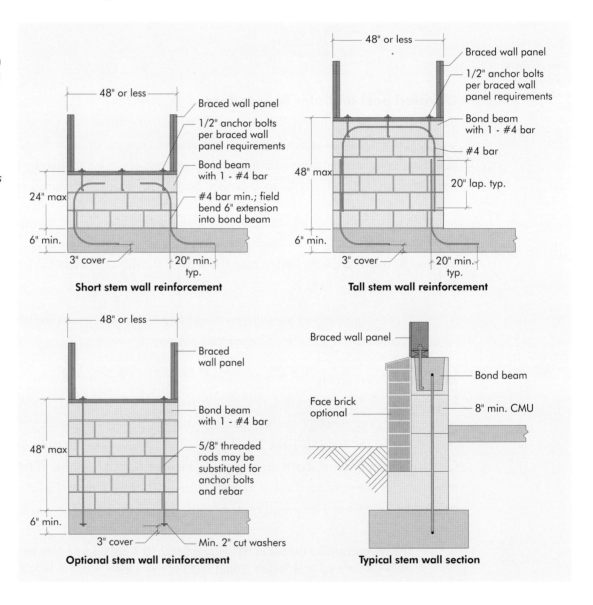

Braced wall panel support for Seismic Design Category D₂

Special bracing support provisions apply to the highest seismic category covered in the IRC. IRC Section R602.10.7.1, regarding braced wall panel support for SDC D_2, is unchanged from the 2006 IRC with one notable exception: the distinction between interior and exterior braced wall lines has been eliminated. Although the section now applies to all braced wall lines, note that the exception applies only to interior braced wall panels/lines. **FIGURE 8.10** illustrates foundation support at a 50-foot interval for a one-story home in SDC D_2. **FIGURE 8.11** illustrates all interior braced wall panels/lines supported for a two-story home in SDC D_2.

FIGURE 8.10

Single-story foundation support in SDC D₂

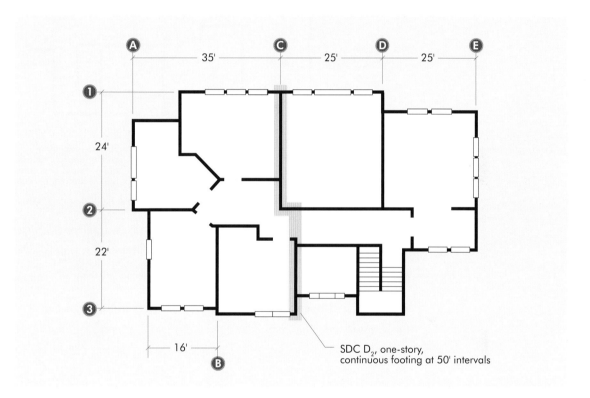

SDC D_2, one-story, continuous footing at 50' intervals

FIGURE 8.11

Two-story foundation support in SDC D₂

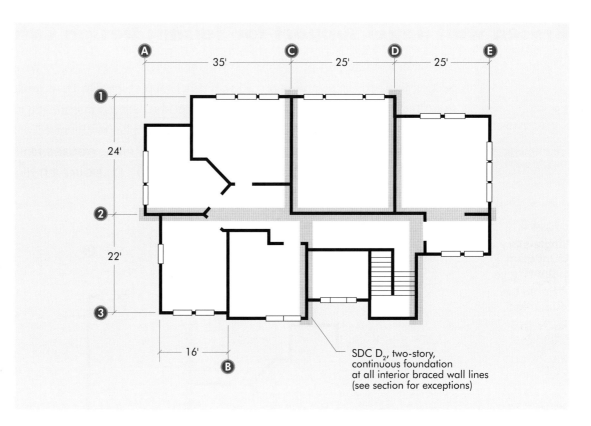

SDC D₂, two-story, continuous foundation at all interior braced wall lines (see section for exceptions)

Note that in the above figure, wall line B was not used as an interior braced wall line. (Not all interior wall lines must be used as interior braced wall lines.) If it had been used as an interior braced wall line, then it would have required a continuous foundation as well.

R602.10.7.1 Braced wall panel support for Seismic Design Category D₂. *In one-story buildings located in Seismic Design Category D₂, braced wall panels shall be supported on continuous foundations at intervals not exceeding 50 feet (15 240 mm). In two story buildings located in Seismic Design Category D₂, all braced wall panels shall be supported on continuous foundations.*

Exception: *Two-story buildings shall be permitted to have interior braced wall panels supported on continuous foundations at intervals not exceeding 50 feet (15 240 mm) provided that:*

1. *The height of cripple walls does not exceed 4 feet (1219 mm).*

2. *First-floor braced wall panels are supported on doubled floor joists, continuous blocking or floor beams.*

3. *The distance between bracing lines does not exceed twice the building width measured parallel to the braced wall line.*

Bracing anchorage to foundation – wind less than 110 mph and Seismic Design Categories A, B and C

When braced wall panels are supported directly by foundations (e.g., turned-down slab edge, thickened slab, masonry or concrete foundation wall), the wood sill plate must be anchored to the foundation with anchor bolts spaced a maximum of 6 feet on center. There must be a minimum of two bolts per sill plate section, with a bolt located not more than 12 inches and not less than 7 bolt diameters (3.5 inches) from each end of the plate section (IRC Section R403.1.6). Bolts should be at least 1/2 inch in diameter and should extend a minimum of 7 inches into the concrete or masonry foundation. A nut and washer is required on each bolt to hold the plate to the foundation. See **FIGURE 8.12**.

FIGURE 8.12

Requirements to connect sill plates to the foundation

IRC Section R403.1.6

FIGURE 8.13

Example of connection hardware required in IRC Section R602.11.1

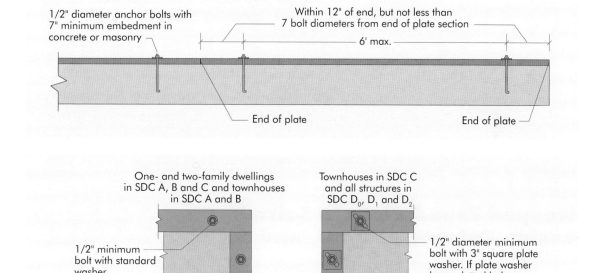

IRC Section R602.11 specifies how braced wall line sill plates are to be anchored to the foundation. IRC Section R403.1.6 specifies the minimum number, size and spacing of anchor bolts.

> **R602.11 Wall anchorage.** *Braced wall line sills shall be anchored to concrete or masonry foundations in accordance with Sections R403.1.6 and R602.11.1.*

For one- and two-family dwellings in SDC A, B and C and townhouses in SDC A and B, braced wall lines are to be connected to the foundation with 1/2-inch bolts using a nut and washer. For higher seismic requirements (townhouses in SDC C and all structures in SDC D_0, D_1 and D_2), braced wall lines are to be connected to the foundation with 1/2-inch bolts using a 3-inch square plate washer and nut, per IRC Section R602.11.1. See **FIGURE 8.13**. The plate washer is permitted to have a slotted hole, which helps to locate the washer over the sill plate. If the plate washer has a slotted hole, a standard cut washer must be used between the plate washer and the nut. The slotted hole is permitted to be 3/16 inch wider than the bolt diameter, and the slot length is limited to a maximum of 1-3/4 inches.

IRC Section R602.11.1 also permits the use of "approved anchor straps" in lieu of anchor bolts, washers and nuts.

Bracing anchorage to foundation - additional requirements for Seismic Design Categories D₀, D₁ and D₂

The following provisions are required in IRC Section R602.11.1, in addition to the requirements for SDC A, B and C as previously described:

- For all buildings in SDC D₀, D₁ and D₂ and townhouses in SDC C, the plate washers are required to be 0.229 inch x 3 inches x 3 inches between the nut and the sill plate except where approved anchor straps are used.

- The use of a diagonally slotted plate washer is permitted. The slot width is required to be equal to or less than 3/16 inch larger than the bolt diameter and not more than 1-3/4 inch long. When a slotted plate washer is used, a standard cut washer must be used between the nut and plate washer. See *FIGURE 8.13*.

In the 2009 IRC, connection information specific to interior braced wall lines has been omitted. This is the result of the reorganization of the bracing section, which included removal of the distinctions between interior braced wall lines and exterior braced wall lines. The ICC Ad Hoc Committee on Wall Bracing determined that the requirements for all braced wall lines were the same regardless of their location within the structure. Therefore, the attachment and support requirements for exterior and interior braced wall lines are now the same.

Stepped Foundations in Seismic Design Categories D₀, D₁ and D₂

Stepped foundation requirements in the 2009 IRC (Section R602.11.2) are essentially unchanged, although the text was edited for clarification and to include references to appropriate sections of the code. Stepped foundations are similar to cripple walls and are discussed in *CHAPTER 9*.

Foundation requirements for braced wall lines

In general, the foundation is the last major component of a structure's vertical and lateral load paths.

The foundation must be designed to distribute the vertical load (the weight of the structure and loads acting on it) into the soil below the foundation (as discussed in *CHAPTER 1*). When properly designed, the load imparted on the soil by the structure must be less than or equal to what the soil can adequately support with minimal settlement. The heavier the structure, the wider the foundation has to be in order to distribute the load into the soil. A poorer quality soil requires a wider footing.

Lateral loads (that result from lateral events, such as earthquakes and high winds) generate two separate forces that must also be distributed into the soil below the foundation: sliding forces and overturning forces (discussed in *CHAPTER 2*). Sliding forces push the structure along the surface of the ground. In structures covered by the IRC, the foundation extends below the surface of the ground and resists sliding forces by a combination of friction against the surrounding soil surface and the compressive resistance of the soil acting against the foundation in the direction of the lateral load. If the structure is adequately connected to the foundation, and the foundation extends below the surface of the ground as specified by the IRC, sliding forces historically have not been a problem.

Overturning forces push – or "tip" – the structure over. An example of this phenomenon is moving a refrigerator across the floor by pushing the top: as long as the refrigerator can roll forward without any resistance, it remains upright. But as soon as the refrigerator wheels hits the carpet, and the bottom stops moving, the force pushing the top of the refrigerator forward may tip it over. A secure foundation, anchored in the soil to prevent sliding forces, resists overturning forces in two ways. First, on the side opposite of the lateral load, the structure's footing is prevented from rotating downward by the soil's resistance. Second, on the side with the acting lateral load, the foundation is prevented from pulling up and out of the ground by the friction of the soil on each side of the foundation, the weight of the soil on top of the footing, the weight of the structure, and the weight of the foundation itself. See *FIGURE 8.14*.

FIGURE 8.14

Actions of soil on foundations

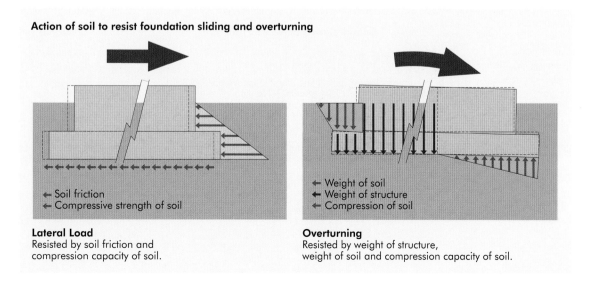

Action of soil to resist foundation sliding and overturning

← Soil friction
← Compressive strength of soil

Lateral Load
Resisted by soil friction and compression capacity of soil.

← Weight of soil
← Weight of structure
← Compression of soil

Overturning
Resisted by weight of structure, weight of soil and compression capacity of soil.

If these upward forces are strong and the structure is well attached to the foundation, the foundation must be very deep and/or possess a wide footing. A wide footing transfers upward forces into a wedge of soil on each side of the footing: if the footing is going to pull up and out, it must lift the wedge of soil with it. Therefore, a wide footing effectively increases the weight of the foundation, thus increasing its resistance to overturning.

Continuous footings

There are a number of foundation types that may be built per IRC Section R104.11. These include piles, piers, floating foundations and discontinuous foundations; however, because the focus of this publication is bracing and the IRC, the content found here is generally limited to continuous footings. Continuous footings, for both interior and exterior braced wall lines, are the only type of foundation prescriptively required (IRC Sections R403.1.2, R403.1.6 and R403.1.6.1) or described (R403.1.3 and R602.10.6) in the IRC.

The IRC restrictions on minimum-footing size and application limit the actual loads on the foundation. The vertical loads are limited by prescriptive requirements in IRC Chapter 3 (R301.2). The lateral loads are limited by the prescriptive requirements in IRC Chapter 4 (R403) and the wall anchorage requirements of IRC Chapter 6 (R602.11.1). See **FIGURE 8.15**. In SDC D_0, D_1 and D_2, additional reinforcement requirements for foundations are specified, as noted in **FIGURE 8.16**.

FIGURE 8.15

Minimum exterior bracing wall requirements

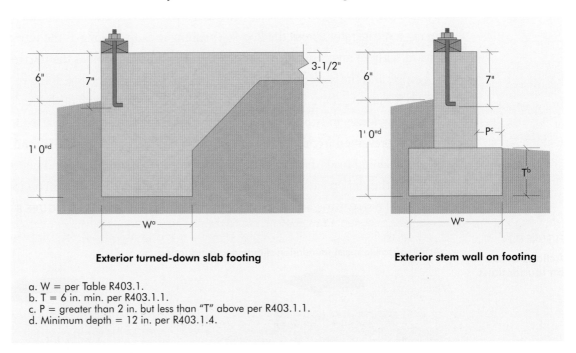

Exterior turned-down slab footing   **Exterior stem wall on footing**

a. W = per Table R403.1.
b. T = 6 in. min. per R403.1.1.
c. P = greater than 2 in. but less than "T" above per R403.1.1.
d. Minimum depth = 12 in. per R403.1.4.

FIGURE 8.16

Bracing wall foundation requirements for SDC D$_0$, D$_1$ and D$_2$

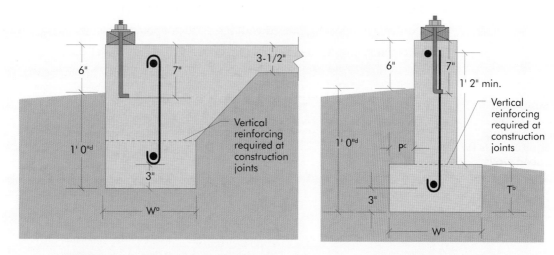

Exterior turned-down slab footing[e,f]   Exterior stem wall on footing[g,h,i]

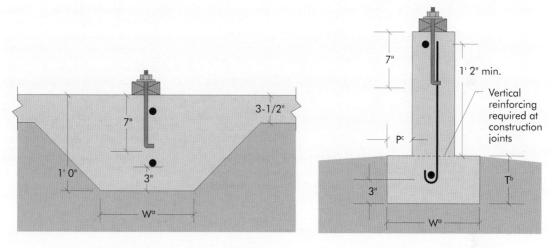

Interior thickened (monolithic) slab footing[e]   Interior stem wall on footing[g,h,i]

a. W = per Table R403.1.

b. T = 6 in. min. per R403.1.1.

c. P = greater than 2 in. but less than "T" above per R403.1.1.

d. Minimum depth = 12 in. per R403.1.4.

e. **Turned-down slab:** Reinforcing (#4 bar top and bottom) shown for SDC D$_0$, D$_1$ and D$_2$. If monolithic – single #5 bar or (2) #4 bars permitted in middle third per R403.1.3.2.

f. **Turned-down slab:** Vertical bars shown for SDC D$_0$, D$_1$ and D$_2$ when construction joint is between footing and stem wall – #4 bars at 4 ft o.c. is recommended.

g. **Stem wall and footing:** Vertical bars shown for SDC D$_0$, D$_1$ and D$_2$ when construction joint is between footing and stem wall – #4 bars at 4 ft o.c. per R403.1.3.

h. **Stem wall and footing:** Vertical bars shown for SDC D$_0$, D$_1$ and D$_2$ when grouted masonry stem wall over concrete footing – #4 bars at 4 ft o.c. per R403.1.3. *While not a code requirement, similar detail for thickened slab when done in two pours is recommended.*

i. **Stem wall and footing:** Reinforcing (#4 bar top and bottom) shown for SDC D$_0$, D$_1$ and D$_2$ per R403.1.3.1.

As can be seen in the many code references above, the requirements for foundations and cripple wall bracing supporting braced wall lines are peppered throughout the IRC. These assorted requirements have been collected and consolidated into **TABLE 8.3**.

TABLE 8.3

Foundation and cripple wall bracing requirements for braced walls

Exterior or Interior	SDC	Total Number of Stories	Applicable Section of IRC	Foundation Requirement	Cripple Wall Bracing Requirements
Exterior	A-D$_1$	1, 2, 3	R403.1	All exterior walls shall be supported on a continuous solid or fully grouted masonry or concrete foundation	IRC Section R602.10.9 (SDCs D$_0$ and D$_1$ must also comply with IRC Section R602.10.9.1)
	D$_2$	1, 2	R602.10.7.1	All exterior braced wall lines must be supported on a continuous foundation	IRC Sections R602.10.9 and R602.10.9.1
Interior	A-C	1, 2, 3	R403.1	Continuous foundation is not required	IRC Section R602.10.9
	D$_0$ & D$_1$	1, 2, 3	R403.1.2	If plan dimension is less than 50', continuous foundation not required under interior braced walls	IRC Sections R602.10.9 and 1.5 times the bracing requirements in IRC Tables R602.10.1.2(1) and (2)
	D$_2$	1	R403.1.2 & R602.10.7.1	If plan dimension is greater than 50', a continuous foundation is required under interior braced walls	IRC Sections R602.10.9 and 1.5 times the bracing requirements in IRC Tables R602.10.1.2(1) and (2)
		2	R602.10.7.1	All interior braced wall lines shall have continuous footings[a]	IRC Tables R602.10.1.2(1) and (2)

a. SDC D$_2$ single-story provisions shall be permitted provided that all of the following conditions are met:
 - cripple wall height less than or equal to 4 ft, and
 - first floor braced wall panels are supported on doubled floor joists, and
 - the distance of braced wall lines does not exceed the building width measured parallel to the braced wall lines

Bracing wall anchorage to foundations

The foundation anchorage requirements for braced wall lines are given in Chapters 4 and 6 of the IRC. **TABLE 8.4** summarizes these requirements. Note that the applicable references in the 2009 IRC are given in in the fourth column of the table.

TABLE 8.4

Foundation anchorage requirements for braced walls

Exterior or Interior	SDC	Total Number of Stories	Applicable Section of 2009 IRC	Anchorage Requirement
Exterior	A-C[a]	1, 2, 3	R403.1.6	1/2" bolt with 7" embedment spaced a maximum of 6' with nut and standard cut washer
Exterior	D_0-D_2	1, 2	R403.1.6.1, Item 1	1/2" bolt with 7" embedment spaced a maximum of 6' with nut and 3" x 3" x 0.229" thick washer
Exterior	D_0-D_2	3	R403.1.6.1, Items 1 & 4	1/2" bolt with 7" embedment spaced a maximum of 4' with nut and 3" x 3" x 0.229" thick washer
Interior	A-C[a]	1, 2, 3	R403.1.6	Where foundation required, 1/2" bolt with 7" embedment spaced a maximum of 6' with nut and standard cut washer
Interior	D_0-D_2	1, 2	R403.1.6.1, Items 1 & 2	1/2" bolt with 7" embedment spaced a maximum of 6' with nut and 3" x 3" x 0.229" thick washer
Interior	D_0-D_2	3	R403.1.6.1, Items 1, 2 & 4	1/2" bolt with 7" embedment spaced a maximum of 4' with nut and 3" x 3" x 0.229" thick washer

a. Townhouses in SDC C are required to have 1/2 in. bolt with 7 in. embedment spaced a maximum of 6 ft with nut and 3 in. x 3 in. x 0.229 in. thick plate washers per IRC Section R403.1.6.1.

Uplift load path

New for the 2009 IRC is a section on the uplift load path. The same wind load that necessitates bracing in a structure also acts to uplift exterior walls that support roof framing. As IRC Section R602.10.1.2.1 is currently written, only braced wall panels on the exterior walls that support roof framing require an uplift load path. Therefore, braced wall panels that only support a floor above are not subject to these uplift requirements. The intention of this section was to provide an adequate load path for the lateral force resisting system of the structure (wall bracing) as a minimum. From an engineering perspective, however, this provision appears to be incomplete because uplift loads are possible at the end of every roof truss or rafter, whether it bears on a bracing panel or not. This is perhaps understandable in that these provisions were proposed by the ICC Ad Hoc Bracing Committee and their mandate covered wall bracing only. To provide for the appropriate attachment of roof framing to the supporting walls for the whole structure, the uplift requirements of Chapter 6 of the IRC must be used in conjunction with the requirements of IRC Section R802.11, discussed below.

The prescriptive nailing provisions in IRC Table R602.3(1), Item 5 specifies two 16d toe nails for the "rafter to plate..." Based on historical performance, this prescriptive schedule is deemed to provide sufficient attachment for 90 mph winds, Exposure B, 5:12 pitch or greater, and a roof span of 32 feet or less. For this reason, IRC Section R602.10.1.2.1, Item 1.1, exempts structures meeting these characteristics from the uplift load path requirements, provided the roof is attached in accordance with IRC Table R603(1), Item 5.

From IRC Section R602.10.1.2.1, Item 1.1, it can be inferred that the IRC Table R602.3(1), Item 5 prescriptive nailing provisions provide 100 pounds of net uplift resistance to properly attached roofs. No additional attachment is required until the uplift loads specified in IRC Table R802.11 exceed this amount. In addition, IRC Section R602.10.1.2.1, Item 1.2 provides the effective weight of each full wall above the wall-to-floor connection in question as 60 plf. This amount can be subtracted from the uplift amount, as the uplift requirement for each lower floor is calculated. (See example on page 200) Given this information, the user is able to use the truss or rafter connection uplift force provided in IRC Table R802.11 and its footnotes to compute the net uplift requirement for a structure at a given location. Note that the footnotes of IRC Table R802.11 are important and must be considered.

When the uplift resistance of the roof framing provided in IRC Table R802.11 is reduced by the number of walls above the connection in question or the location in the structure, and the net result is less than 100 plf, the prescriptive nailing requirements of IRC Table R603(1) are deemed to be sufficient to provide an adequate load path.

When the uplift resistance of the roof framing provided in IRC Table R802.11 is reduced by the number of walls above the connection in question or the location in the structure, and the net result is greater than 100 plf, the prescriptive nailing requirements of IRC Table R602.3(1) are insufficient to provide an adequate load path and the following options are available:

1. Installation of an approved uplift framing anchor of sufficient capacity to resist the net uplift force.

2. IRC Section R104.11 permits engineering design to be used to determine other nailing schedules or details that may provide sufficient uplift resistance. When using the engineering design option, the bracing and uplift capacity of the braced wall panel can be accomplished using the combined shear and uplift capacity of the braced wall sheathing panel. (This information is provided in the American Wood Council's *Special Design Provisions for Wind and Seismic*, an IRC-recognized design guide that contains related design information for wood structural panels.)

3. Engineering analysis can be used to determine the wind uplift capacity in lieu of IRC Table R802.11. Such analysis can take into account heavier roofing materials and other details not accounted for in the tables.

4. Other referenced documents, such as the American Wood Council's *Wood Frame Construction Manual,* can be used to generate uplift requirements and prescriptive hold-down requirements.

This new section of the IRC is provided below:

> **R602.10.1.2.1 Braced wall panel uplift load path.** *Braced wall panels located at exterior walls that support roof rafters or trusses (including stories below top story) shall have the framing members connected in accordance with one of the following:*
>
> **1.** *Fastening in accordance with Table R602.3(1) where:*
>
> > **1.1** *The basic wind speed does not exceed 90 mph (40 m/s), the wind exposure category is B, the roof pitch is 5:12 or greater, and the roof span is 32 feet (9754 mm) or less, or*
> >
> > **1.2** *The net uplift value at the top of a wall does not exceed 100 plf (146 N/mm). The net uplift value shall be determined in accordance with Section R802.11 and shall be permitted to be reduced by 60 plf (86 N/mm) for each full wall above.*
>
> **2.** *Where the net uplift value at the top of a wall exceeds 100 plf (146 N/mm), installing approved uplift framing connectors to provide a continuous load path from the top of the wall to the foundation. The net uplift value shall be as determined in Item 1.2 above.*
>
> **3.** *Bracing and fasteners designed in accordance with accepted engineering practice to resist combined uplift and shear forces.*

Example: What is the net uplift capacity for each level of a two story home located in a 110 mph Exposure B wind zone? The width of the house is 32 feet and none of the footnotes to IRC Table R802.11 apply. The house has no roof overhangs and the roof framing is at 24 inches on center.

Solution: From IRC Table R802.11, the wind uplift acting at the roof-to-second story wall is −467 pounds per connection. The negative sign indicates that it is an uplift load.

- Capacity of the roof-to-second story wall connection – There is no wall above this connection, so the uplift at this point is 467 pounds per connection. IRC Table R602.33(1), Item 5 attachment schedule provides 100 pounds per connector of uplift resistance. If this attachment schedule is used, a connector with a minimum capacity of (467 pounds connector – 100 pounds per connector =) 367 pounds per connector is required. If this attachment schedule is not used, a connector with the full 467 pound capacity is required.

- Capacity of the second story-to-first story connection – As each lineal foot of wall contributes 60 pounds of uplift resistance, and there are 2 feet of wall per connector (roof framing/connectors on 24-inch centers), the uplift requirement for the second story-to-first story connection will be (467 – (2 x 60) =) 347 pounds per connector.

- Capacity of the first story-to-foundation connection – In this case, there are two stories above the connection, and each lineal foot of wall contributes 60 pounds of uplift resistance, and there are 2 feet of wall per connector (roof framing/connectors on 24-inch centers). The uplift requirement for the first story-to-foundation connection will be (467 – (2 x 2 x 60) =) 227 pounds per connector.

While not specified in the IRC, it is reasonable to assume that IRC Table R602.3(1) nailing schedules provide at least some of this uplift resistance at the second-to-first and first-to-foundation locations. The use of 100 pounds per foot for the capacity of the IRC Table R602.3(1) connections at these two locations is probably as reasonable as the assumption that IRC Table R602.3(1), Items 1-5 provide 100 plf uplift resistance capacity. From a historical perspective, these second-to-first-floor and first-floor-to-foundation connections specified have been as effective as the roof-to-floor connections. As noted elsewhere, deviations from the code provisions should be approved by the building official.

CHAPTER 9

Cripple Walls

What is a cripple wall?

A cripple wall is a less-than-full-height wall that is used to raise the elevation of a floor above the foundation, as shown in *FIGURE 9.1*. Cripple walls are also often used in conjunction with a stepped foundation to maintain a common plate elevation when the foundation drops away from the plate line, accommodating a sloped building site, as shown in *FIGURE 9.2*. In either case, a cripple wall has the same limitations as any other stud wall in that it has no lateral load capacity without bracing.

FIGURE 9.1

**Cripple Wall
used to raise
floor elevation**

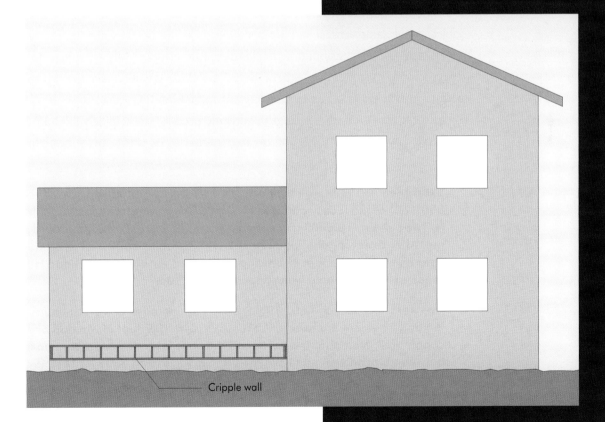

Cripple wall

FIGURE 9.2

Cripple Wall used with a stepped foundation

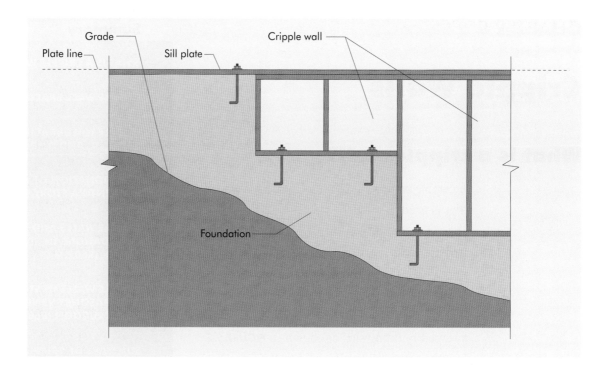

Cripple wall bracing

Even though they are less than full height, cripple walls are still a part of the load path and are subject to the same vertical and horizontal loads as full-height walls. In fact, because gravity and lateral loads get larger as building height increases, cripple walls are subject to even greater loads than the walls in stories above them. Because of these greater loads, the IRC (Section R602.10.9) permits cripple walls to be braced using any method in IRC Tables R602.10.1.2(1) and (2) in any SDC except D_2, as long as the length of bracing used for the cripple wall is 15 percent greater (multiplied by a factor of 1.15) and the maximum braced wall panel spacing in the *cripple wall line* is reduced from 25 to 18 feet.

> **Example:** A cripple wall supports the first story of a two-story residence. Per Table R301.2(1) of the locally adopted IRC, the design wind speed is 90 mph and the SDC is A. The builder is planning on using Method SFB (structural fiberboard sheathing) bracing. After reviewing the requirements of IRC Tables R602.10.1.2(1) and (2), the builder determines that 9.5 feet of bracing is required for the bottom floor. How much bracing is required for the cripple wall supporting that floor?
>
> **Answer:** IRC Section R602.10.9 requires an increase in the minimum amount of bracing by 15 percent; therefore, 9.5 feet x 1.15 = 10.9 feet of Method SFB bracing for the cripple wall. Note that this section of the code also decreases the maximum distance between braced wall panels <u>in the cripple walls</u> from 25 to 18 feet. (See *CHAPTER 3* for clarification on measuring distance between braced wall lines.)

Cripple wall bracing in Seismic Design Categories D₀, D₁ and D₂

As the seismic force on a structure increases, the lateral load acting on the cripple walls also increases. IRC Section R602.10.7.1 (discussed above) requires continuous foundations in SDC D_2, and only at intervals not to exceed 50 feet.

In SDC D_0, D_1 and D_2, IRC Section R602.10.9.1 requires increased *exterior* cripple wall bracing when *interior* braced walls are not supported by a continuous foundation. (Note that this requirement is in addition to the requirements of IRC Section R602.10.9, discussed above.) Without direct interior foundation support, these interior braced wall lines transfer the lateral forces into the floor diaphragm instead of the foundation. The floor diaphragm, in turn, transfers these lateral forces into the exterior cripple walls. Therefore, the wall bracing at the exterior cripple walls (specified in IRC Table R602.10.1.2(2)) must be increased to accommodate the increased load. This is done by multiplying the length of the required exterior cripple wall bracing parallel to the unsupported interior braced wall by a factor of 1.5, as shown in **FIGURE 9.3**.

FIGURE 9.3

Cripple wall bracing in SDC D₀, D₁ and D₂

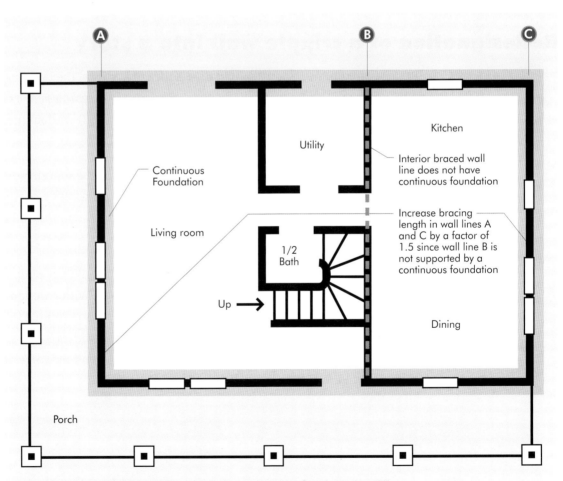

If interior braced wall lines are not supported by a continuous foundation (i.e. "B"); Increase exterior cripple wall bracing by a factor of 1.5 (i.e., braced wall lines A & C).

If Method WSP (wood structural panel) bracing is used and the length of wall bracing cannot be increased by a factor of 1.5 because the wall line is not long enough to accommodate the increased bracing length, the nail spacing along the perimeter of the Method WSP sheathing may be reduced from 6 inches to 4 inches on center, per IRC Section R602.10.9.1. Note that in this case it is not necessary to also increase the length of bracing by the 1.5 factor; the provision recognizes that the shear capacity of wood structural panels (Method WSP) is increased sufficiently by increasing the perimeter nailing.

The code does not stipulate what to do if any of the other bracing methods (Methods DWB, SFB, PBS, PCP or HPS) are used for cripple wall bracing and there is insufficient length within the braced wall line to permit a 50 percent increase in bracing length. In such a case, an engineered design or other alternative bracing approved by the building official must be incorporated. To avoid an engineered design, the obvious solution is to use Method WSP bracing with 4-inch on center perimeter nail spacing as discussed above. Mixing bracing methods from story to story is permitted in IRC Section R602.10.1.1, Item 1; therefore, a bracing method substitution can be confined to only the cripple wall.

Redesignation of a cripple wall into a story

Per IRC Section R602.10.9.2, in all wind or seismic applications, the building designer is permitted to redesignate/redefine the cripple wall as a story, regardless of its height, and then use the bracing requirements of IRC Tables R602.10.1.2(1) or (2) without the cripple wall adjustments required by IRC Sections R602.10.9 and R602.10.9.1.

When redesignation is used, the cripple walls are considered the first story, while the first full-height story becomes the second story, etc., as shown in **FIGURES 9.4a** through **9.4d**. Redesignation is optional (*"shall be permitted"* in code language) and the builder in any SDC may choose to exercise this option.

Note that in SDC D_2, IRC Table R602.10.1.2(2) only permits buildings of up to two stories, thus limiting the redesignation option to one-story buildings. With redesignation, a one-story becomes a two-story, which is the maximum number of stories permitted in SDC D_2. If a two-story structure is located above a redesignated cripple wall foundation, the second story would become the third story. A three-story structure in SDC D_2 is required to be designed in accordance with the International Building Code.

FIGURE 9.4a

Redesignating cripple walls

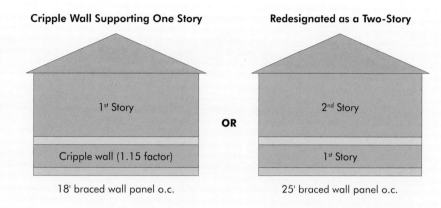

FIGURE 9.4b

<110 mph wind and SDC A-D₁ example

IRC Section R602.10.9

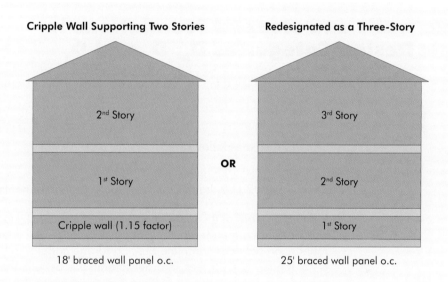

FIGURE 9.4c

<110 mph and SDC D₂ example

IRC Section R602.10.9

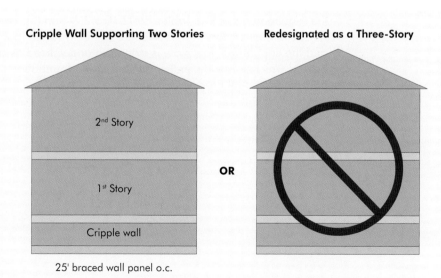

FIGURE 9.4d

<110 mph and SDC D₂ example

IRC Section R602.10.2

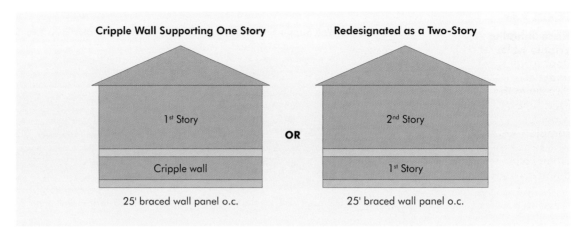

Cripple walls and stepped foundations in Seismic Design Categories D₀, D₁ and D₂

The provisions in the 2009 IRC that address cripple walls and stepped foundations have been rewritten for clarity, but the basic requirements remain unchanged. When a cripple wall in SDC D_0, D_1 and D_2 is used in a stepped foundation with a vertical step exceeding 4 feet, the cripple wall must be constructed as follows:

1. IRC Section R602.11.2, Item 1 contains multiple requirements:

- If 8 feet or more of the lowest floor is anchored directly to the foundation (the cripple wall does not run under this 8-foot portion of the wall line) as shown in **FIGURE 9.5**, the foundation provides complete bracing for the cripple wall, as long as the wall above is not longer than 25 feet. In order to gain this benefit, the cripple wall top plate must extend over the foundation at least 4 feet and be anchored with at least two bolts, as shown in **FIGURE 9.5**. Note that that first story wall must be braced in accordance with the requirements of IRC Tables R602.10.1.2(1) and (2). The intent of the code provision is that, if the first story wall is anchored directly to the foundation for at least 8 feet and the cripple wall top plate is extended and spliced as directed, then the first story braced wall line is sufficiently attached to the foundation to complete the load path and no additional bracing is required.

FIGURE 9.5

8 lineal feet or greater of lower-floor framing attached directly to the foundation provides complete cripple wall bracing when splice detail is used

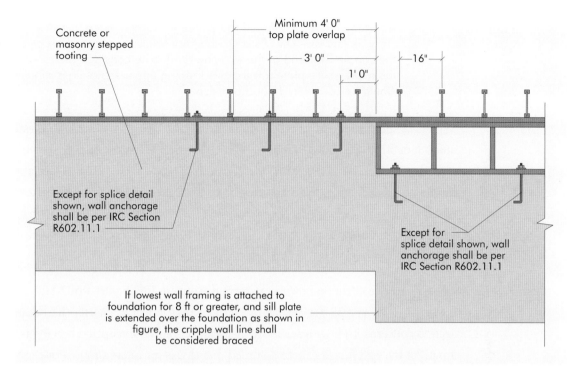

Concrete or masonry stepped footing

Minimum 4' 0" top plate overlap

3' 0"

1' 0"

16"

Except for splice detail shown, wall anchorage shall be per IRC Section R602.11.1

Except for splice detail shown, wall anchorage shall be per IRC Section R602.11.1

If lowest wall framing is attached to foundation for 8 ft or greater, and sill plate is extended over the foundation as shown in figure, the cripple wall line shall be considered braced

- As shown in **FIGURE 9.6**, a metal tie (IRC Figure R602.11.2) can be used as an alternative to the cripple wall top plate's 4-foot overlap of the foundation, if distance "A" is greater than 8 feet. No additional wall bracing is required in the cripple wall if the splice is properly made.

FIGURE 9.6

Stepped foundation construction

IRC Figure R602.11.2

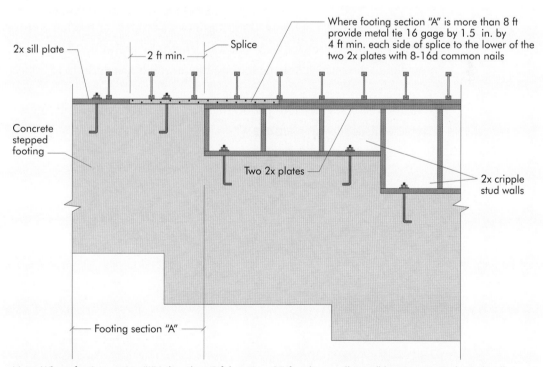

2x sill plate

2 ft min.

Splice

Where footing section "A" is more than 8 ft provide metal tie 16 gage by 1.5 in. by 4 ft min. each side of splice to the lower of the two 2x plates with 8-16d common nails

Concrete stepped footing

Two 2x plates

2x cripple stud walls

Footing section "A"

Note: Where footing section "A" is less than 8 ft long in a 25-foot-long wall, install bracing at cripple stud wall.

- When the distance "A" is less than 8 feet in a 25-foot-long wall, there is insufficient attachment directly into the foundation to complete the lateral load path for the first-story braced wall line. As such, the total required length of bracing must be applied to the cripple-wall-end of the wall line for it to be considered braced. In this event, neither the metal splice plate nor the overlapping upper top plate is required, although either would be considered good practice.

While not specifically addressed in IRC Section R602.11.2, it is reasonable to assume that the amount of bracing required for a wall line above a stepped foundation as shown in **FIGURE 9.6** is in accordance with IRC Tables R602.10.1.2(1) and (2). The cripple wall that supports a portion of the braced wall line above is subject to the 15 percent increase (1.15 times the bracing length required) and braced wall panel spacing requirements of IRC Section R602.10.9. Note that in SDC D_2, when using the seismic bracing table (IRC Table R602.10.1.2(2)), the length of the braced wall line above the stepped foundation is used for determining the bracing length of the cripple wall.

The code is also silent on how to interpret the note in IRC Figure R602.11.2, *"When the distance "A" is less than 8 feet in a 25-foot-long wall..."*, if the wall is longer or shorter than than 25 feet. Since this section concerns itself with seismic considerations and the required bracing is proportional to the length of the wall line, it can be assumed that this is meant as a ratio. For example, if the wall were 50 feet long, then distance "A" would be interpreted as being increased to "less than 16 feet."

$$A = L \times (8/25)$$
Where: A = Distance "A" in IRC Figure R602.11.2 (feet)
L = length of braced wall line (feet)

In other words, 8 feet of foundation length (with proper connections) provides all the seismic bracing that is required for a 25-foot wall length. By proportion, 16 feet of foundation length provides all the seismic bracing required for a wall line length up to 50 feet.

Note also that a stepped foundation shall not be constructed by placing a cripple wall over a cripple wall, as shown in **FIGURE 9.7**.

FIGURE 9.7

FIGURE 9.7

Cripple walls shall not be placed one over the other

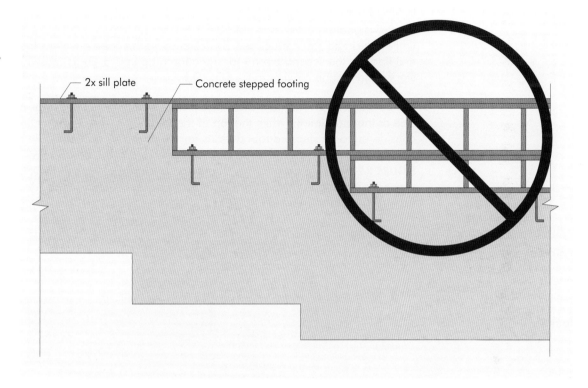

2x sill plate

Concrete stepped footing

2. IRC Section R602.11.2, Item 2 clarifies that when a cripple wall extends over a stepped footing in SDC D_0, D_1 and D_2 for the full length of the wall line, the bracing requirements of a story shall be applied to that cripple wall, as shown in **FIGURE 9.8**. Therefore, the cripple-wall-specific requirements of IRC Sections R602.10.9 and R602.10.9.1 are required. If any single wall line of a building has a full-length cripple wall over a stepped foundation, as illustrated in **FIGURE 9.8**, all cripple walls above the stepped foundation have to meet the story requirements specified in IRC Section R602.11.2.

FIGURE 9.8

Cripple wall in SDC D_0, D_1 and D_2, treated as a story

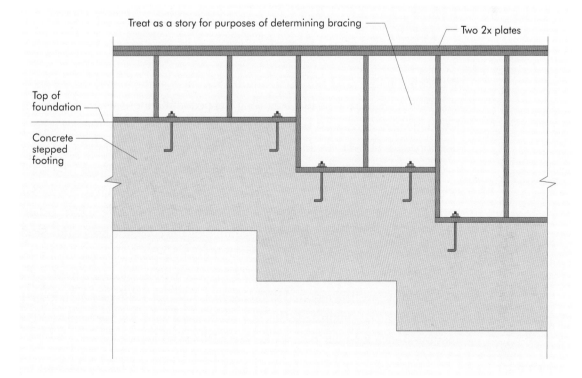

Treat as a story for purposes of determining bracing

Two 2x plates

Top of foundation

Concrete stepped footing

In SDC D$_2$, no more than two stories are permitted (IRC Table R602.10.1.2(2)). As such, the redesignation provisions of R602.10.9.2 for buildings with stepped foundations are limited to single-story structures. By designating the cripple wall as the first floor, and the first floor as the second floor, redesignation redefines a single-story building as a two-story building (the maximum for SDC D$_2$).

3. IRC Section R602.11.2, Item 3 clarifies that foundations having steps only on the bottom side of the foundation (while the top of the footing has a common elevation) are not considered to be stepped foundations for the purposes of the IRC. In these cases, normal foundation attachment requirements (IRC Sections R403.1.6 and R602.11.1) apply. See **FIGURE 9.9**. The reference to the foundation anchorage section, IRC Section R403.1.6, was added to the 2009 IRC for clarity. Plate washer requirements for SDC D$_0$, D$_1$ and D$_2$ are provided in IRC Section R602.11.1.

FIGURE 9.9

Foundation in SDC D$_0$, D$_1$ and D$_2$ not considered to be a stepped footing

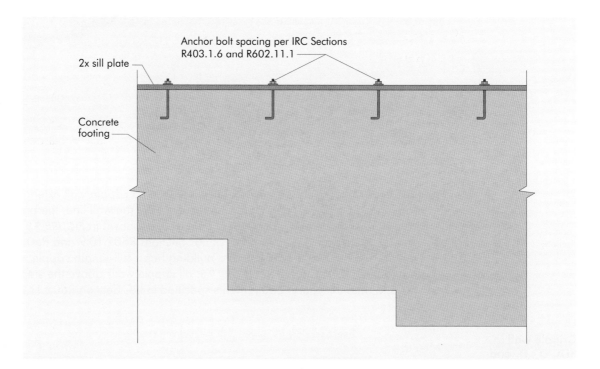

In SDC D$_0$, no more than two stories are permitted.

Anchor bolt spacing per IRC Sections R403.1.6 and R602.11.1

2x sill plate

Concrete footing

While the provisions of IRC Section R602.11.2 are specific to SDC D$_0$, D$_1$ and D$_2$, this is the only instruction offered by the IRC regarding stepped foundations when the elevation of the top of the foundation and the lowest step varies by more than 4 feet. Even though these provisions were developed for SDC D$_0$, D$_1$ and D$_2$, they provide guidance for cripple walls used in stepped foundations in other Seismic Design Categories and high wind speed areas. The bracing provisions themselves are less demanding for lower Seismic Design Categories (and vary by wind pressure) so it can be assumed that the provisions of IRC Section R602.11.2 would be conservative when applied to SDC A, B and C. While this is not required by the code, the use of IRC Section R602.11.2 in SDC A, B and C is encouraged when structures do not have the benefit of a design professional to detail such situations. If IRC Section R602.11.2 is used for guidance in SDC A and B, the 3-inch x 3-inch plate washers required in IRC Section R602.11.1, referenced in Item 3 of IRC Section R602.11.2, can be eliminated. This is in accordance with IRC Section R403.1.6.1, which only requires 3-inch x 3-inch plate washers for SDC C and higher.

Cripple walls in stone and masonry veneer construction

In accordance with the new 2009 IRC Section R602.12, and IRC Table R602.12(1), where stone or masonry veneer is installed in SDC A, B or C, the amount of bracing in all wall lines with one or more stories above shall be per the bracing requirements of IRC Section R602.10.1.2(2), as modified by IRC Table R602.12(1). Note that these provisions apply to townhouses only, as one- and two-family dwellings are exempt from the bracing provisions for SDC A, B and C, per IRC Section R301.2.2, Exception.

In SDC C, the bracing amounts required by IRC Table R602.10.1.2(2) for the bottom story of a two-story or bottom and middle story of a three-story are multiplied by 1.5. In such cases, because cripple wall bracing must be multiplied by 1.15 times the required bracing in the wall above (IRC Section R602.10.9), the bracing required for a cripple wall supporting a stone or masonry veneered wall above is (1.15 x 1.5 =) 1.725 times the bracing requirements of R602.10.1.2(2).

Cripple walls are not permitted in SDC D_0, D_1 and D_2 where stone or masonry veneer is installed, per IRC Section R602.12 and IRC Table R602.12(2), Footnote a. Note that the amount of wind bracing in IRC Table R602.10.1.2(1) is unaffected by the installation of stone and masonry veneer.

Summary of cripple wall provisions

TABLE 9.1 summarizes the most commonly used cripple wall bracing provisions.

TABLE 9.1

Cripple wall bracing – wind and seismic (based on SDC)

Number of Stories	SDC & Wind Speed	Cripple Wall Bracing
1	SDC A–D$_1$ & Wind <110 mph	Brace same method as wall above but add 15% to the bracing length required by IRC Tables R602.10.1.2(1) and R602.10.1.2(2), and reduce spacing of braced wall panel to 18 ft o.c. or Redesignate cripple wall as first story, per R602.10.9.2, and brace all walls as stories with a braced wall panel spacing of 25 ft o.c.
	SDC D$_2$ & Wind <110 mph	Brace as required by IRC Tables R602.10.1.2(1) and R602.10.2.2.(2), and reduce spacing of braced wall panels to 18 ft o.c. For interior braced wall lines without a continuous foundation below, parallel exterior cripple wall lengths per IRC Tables R602.10.1.2(1) and R602.10.1.2(2) must be increased by 50%.
2	SDC A–D$_1$ & Wind <110 mph	Brace same method as wall above but add 15% to the bracing length required by IRC Tables R602.10.1.2(1) and R602.10.1.2(2), and reduce spacing of braced wall panel to 18 ft o.c. or Redesignate cripple wall as first story, per R602.10.9.2, and brace all walls as stories with a braced wall panel spacing of 25 ft o.c.
	SDC D$_2$ & Wind <110 mph	Brace as required by IRC Tables R602.10.1.2(1) and R602.10.1.2(2), and reduce spacing of braced wall panel to 18 ft o.c. For interior braced wall lines without a continuous foundation below, parallel exterior cripple wall lengths per IRC Tables R602.10.1.2(1) and R602.10.1.2(2) must be increased by 50%. **Note:** A two-story with a cripple wall foundation cannot be redesignated as a three-story because a three-story structure in SDC D$_2$ is not permitted in per IRC Table R602.10.1.2(2).
3	SDC A–D$_1$ & Wind <110 mph	Brace same method as wall above but add 15% to bracing required by IRC Tables R602.10.1.2(1) and R602.10.1.2(2), and reduce spacing of braced wall panel to 18 ft o.c. **Note:** A three-story with a cripple wall foundation cannot be redesignated as a four-story because a four-story structure is not within the scope of the International Residential Code.
	SDC D$_2$ & Wind <110 mph	Neither a three-story with a cripple wall foundation nor a four-story structure in SDC D$_2$ is within the scope of the International Residential Code.

For cripple walls of buildings with stone or masonry veneer:

- The bracing provisions of IRC Tables R602.12(1) and R602.12(2) apply.

- For multi-story townhouses in SDC C, the bracing length is 1.50 times the amount of bracing required by IRC Table R602.10.1.2(2). See IRC Table R602.12(1).

- For one- and two-family detached dwellings in SDCs D$_0$, D$_1$ or D$_2$, cripple walls are not permitted per IRC Section R602.12.

CHAPTER 10

Wall Bracing for Stone and Masonry Veneer

The 2009 IRC and its earlier editions have included provisions for wall bracing when stone or masonry veneer is installed as a wall covering. These provisions apply only to stone and masonry veneer, and not other wall coverings, because stone and masonry are heavy compared to other cladding materials. When ground motion from an earthquake puts the cladding in motion, the seismic weight (mass) of a cladding material produces load on the building's lateral-load resisting structural elements. In the IRC, bracing panels are the structural elements that resist the motion of the heavy cladding. The seismic lateral load on a building increases with cladding weight and ground motion intensity; therefore, the requirements for this provision are more stringent in higher Seismic Design Categories.

In the past, these provisions could be found in Chapter 7 of the IRC (the wall covering chapter). Because they were located separately from the other bracing provisions (found in Chapter 6 of the IRC), they were sometimes overlooked. For 2009, the stone and masonry veneer bracing provisions have been relocated to the same chapter as the other bracing provisions.

Summary

Bracing provisions for stone and masonry veneer

This chapter provides a discussion of IRC Section R602.12 and related subsections that address stone and masonry veneer. A summary of these provisions is provided in **TABLES 10.3** and **10.4**.

> **R602.12 Wall bracing and stone and masonry veneer.** *Where stone and masonry veneer is installed in accordance with Section R703.7, wall bracing shall comply with this section.*
>
> *For **all buildings** in Seismic Design Categories A, B and C, wall bracing at exterior and interior braced wall lines shall be in accordance with Section R602.10 and the additional requirements of Table R602.12(1).*

According to this excerpt of the code, *all buildings* in SDC A, B and C are subject to the requirements of IRC Table R602.12(1); however, IRC Section R301.2.2 overrides R602.12 to exempt one- and two-family dwellings in SDC A, B and C and townhouses in SDC A and B. Therefore, the stone and masonry veneer wall bracing requirements provided in IRC Table R602.12(1) (shown in **TABLE 10.1**) are only applicable to townhouses in SDC C.

TABLE 10.1

Braced wall line factor for townhouses in SDC C

IRC Table R602.12(1) Stone or masonry veneer wall bracing requirements, wood or steel framing, seismic design categories A, B and C

Seismic Design Category	Number of Wood Framed Stories	Wood Framed Story	Minimum Sheathing Amount (Length of Braced Wall Line Length)[a]
A or B	1, 2 or 3	all	Table R602.10.1.2(2)
C	1	1 only	Table R602.10.1.2(2)
	2	top	Table R602.10.1.2(2)
		bottom	1.5 times length required by Table R602.10.1.2(2)
	3	top	Table R602.10.1.2(2)
		middle	1.5 times length required by Table R602.10.1.2(2)
		bottom	1.5 times length required by Table R602.10.1.2(2)

a. Applies to exterior and interior braced wall lines.

Note that IRC Table R602.10.1.2(2), referenced in **TABLE 10.1**, is the seismic bracing table discussed in **CHAPTER 6**.

> **R602.12 Wall bracing and stone and masonry veneer.** *...For detached one- or two-family dwellings in Seismic Design Categories D_0, D_1 and D_2 wall bracing and hold downs at exterior and interior braced wall lines shall be in accordance with Sections R602.10 and R602.11 and the additional requirements of Section R602.12.1 and Table R602.12(2). In Seismic Design Categories D_0, D_1 and D_2, cripple walls shall not be permitted, and required interior braced wall lines shall be supported on continuous foundations.*

The third paragraph of IRC Section R602.12 permits stone and masonry veneer to be used in SDC D_0, D_1 and D_2 for detached one- and two-family dwellings only. This is consistent with IRC Section R301.2.2.2.5, Item 7, Exception, which permits the use of masonry veneer as permitted elsewhere in this code. This exception specifies "masonry" veneer only; however, the intent of the code is to permit the use of both stone and masonry veneer, as the two are grouped together elsewhere in the code. For example, the title of Section R703.7 is "Stone and masonry veneer."

Since IRC Section R602.12 does not extend "permission" for the installation of stone and masonry veneer on townhomes in SDC D_0, D_1 and D_2, townhouses would be categorized as "irregular" under IRC Section R301.2.2.2.5, Item 7. The amount of bracing required in percent of the braced wall line length for one- and two-family dwellings that have stone and masonry veneer is provided in IRC Table R602.12(2), reproduced in **TABLE 10.2**.

TABLE 10.2

Braced wall line factors for one- and two- family dwellings in SDC D_0, D_1 and D_2

IRC Table R602.12(2) (Corrected for proper column 4 heading) stone or masonry veneer wall bracing requirements, one- and two- family detached dwellings, seismic design categories D_0, D_1 and D_2

Seismic Design Category	Number of Stories[a]	Story	Minimum Sheathing Amount (Percent of Braced Wall Line Length)[b]	Minimum Sheathing Thickness and Fastening	Single Story Hold Down Force (lb)[c]	Cumulative Hold Down Force (lb)[d]
D_0	1	1 only	35	7/16-inch wood structural panel sheathing with 8d common nails spaced at 4 inches o.c. at panel edges, 12 inches o.c. at intermediate supports. 8d common nails at 4 inches o.c. at braced wall panel end posts with hold down attached	N/A	—
	2	top	35		1900	—
		bottom	45		3200	5100
	3	top	40		1900	—
		middle	45		3500	5400
		bottom	60		3500	8900
D_1	1	1 only	45		2100	—
	2	top	45		2100	—
		bottom	45		3700	5800
	3	top	45		2100	—
		middle	45		3700	5800
		bottom	60		3700	9500
D_2	1	1 only	55		2300	—
	2	top	55		2300	—
		bottom	55		3900	6200

For SI: 1 inch = 25.4 mm, 1 foot = 304.8 mm, 1 pound per square foot = 0.479 kPa, 1 pound-force = 4.448 N.

a. Cripple walls are not permitted in Seismic Design Categories D_0, D_1 and D_2.

b. Applies to exterior and interior braced wall lines.

c. Hold down force is minimum allowable stress design load for connector providing uplift tie from wall framing at end of braced wall panel at the noted story to wall framing at end of braced wall panel at the story below, or to foundation or foundation wall. Use single story hold down force where edges of braced wall panels do not align; a continuous load path to the foundation shall be maintained. [See Figure R602.12.]

d. Where hold down connectors from stories above align with stories below, use cumulative hold down force to size middle and bottom story hold down connectors. [See Figure R602.12.]

In the first printing of the 2009 IRC, the heading for Column 4 of Table R602.12(2) is incorrectly labeled as "Minimum Sheathing Amount (Length of Braced Wall Line Length in Feet)". The correct heading, shown in *TABLE 10.2*, is "Minimum Sheathing Amount (Percent of Braced Wall Line Length)".

Additional requirements for one- and two-family dwellings apply to both the interior and exterior braced wall lines in SDC D_0, D_1 and D_2 (Footnote b to IRC Table R602.12(2)). These requirements define specific bracing materials (wood structural panels) and mechanical hold-down devices at the ends of each braced wall panel. Hold down location and capacity requirements are provided in *TABLE 10.2* and *FIGURE 10.1*, respectively.

In buildings with plan dimensions greater than 50 feet, all interior braced wall lines are to be supported by continuous footings, per IRC Section R403.1.2. Additional requirements are provided in IRC Section R602.10.7.1 and are outlined in *TABLE 8.3*. However, the use of stone and masonry veneer in SDC D_0, D_1 and D_2 requires that *all* interior braced wall lines be supported on continuous foundations regardless of plan dimension. In addition, cripple walls are not permitted to be used with stone and masonry veneer in SDC D_0, D_1 and D_2.

FIGURE 10.1

Hold down location and capacity requirements

IRC Figure R602.12 hold downs at exterior and interior braced wall panels

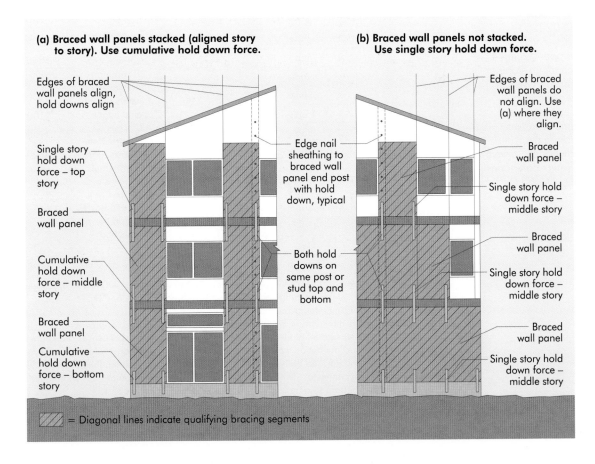

(a) Braced wall panels stacked (aligned story to story). Use cumulative hold down force.

(b) Braced wall panels not stacked. Use single story hold down force.

Edges of braced wall panels align, hold downs align

Single story hold down force – top story

Braced wall panel

Cumulative hold down force – middle story

Braced wall panel

Cumulative hold down force – bottom story

Edge nail sheathing to braced wall panel end post with hold down, typical

Both hold downs on same post or stud top and bottom

Edges of braced wall panels do not align. Use (a) where they align.

Braced wall panel

Single story hold down force – middle story

Braced wall panel

Single story hold down force – middle story

Braced wall panel

Single story hold down force – middle story

⬦ = Diagonal lines indicate qualifying bracing segments

R602.12.1 Seismic Design Categories D$_0$, D$_1$ and D$_2$. *Wall bracing where stone and masonry veneer exceeds the first story height in Seismic Design Categories D$_0$, D$_1$ and D$_2$ shall conform to the requirements of Section R602.10 and R602.11 and the following requirements.*

Although the wording of this section is somewhat confusing, the intent is to permit full-height or less stone or masonry veneer on first story walls without the additional bracing requirements of IRC Table R602.12(2). This provision accommodates short to single-story full-height stone or masonry walls, as often seen on one or two sides of structures. Such walls are thought to resist seismic forces independently from the braced walls to which they are attached. Once these wall veneers exceed full-height, however, the additional seismic weight of the veneer is subject to the additional bracing requirements of IRC Section R602.12.1.

R602.12.1.1 Length of bracing. *The length of bracing along each braced wall line shall be in accordance with Table R602.12(2).*

The minimum bracing length is determined by IRC Table R602.12(2) and is given as a percentage of the braced wall line length. This minimum bracing length is required to resist the increased loads produced by seismic forces resulting from the stone and masonry veneer. IRC Table R602.12(2) supersedes the use of IRC Table R602.10.1.2(2) when stone or masonry veneer exceeds the first story height. Note that the wind bracing requirements of IRC Table R602.10.1.2(1) must also be considered, although it is unlikely that the wind will control in higher seismic areas.

R602.12.1.2 Braced wall panel location. *Braced wall panels shall begin no more than 8 feet from each end of a braced wall line and shall be spaced a maximum of 25 feet on center.*

When stone and masonry veneer extends above the first story wall height, this code section parallels the high-seismic braced wall panel end distance allowance of 8 feet for all stories fully or partially clad with stone or masonry veneer. This requirement applies to one- and two-family homes in SDC D$_0$, D$_1$ and D$_2$.

R602.12.1.3 Braced wall panel construction. *Braced wall panels shall be constructed of sheathing with a thickness of not less than 7/16 inch nailed with 8d common nails spaced 4 inches on center at all panel edges and 12 inches on center at intermediate supports. The end of each braced wall panel shall have a hold-down device in accordance with Table R602.12(2) installed at each end. Size, height and spacing of wood studs shall be in accordance with Table R602.3(5).*

While the code provision itself uses the word "sheathing" in a generic sense, it is clear from the era of the original provision, the increased nail spacing required by the section, and IRC Table R602.12(2), that "sheathing" means wood structural panel sheathing (guidance for other products is not provided). So, for the purposes of this discussion, the term "wood structural panel sheathing" is used for clarity. It is important to note that wood structural panel sheathing is not the same as Method WSP bracing. In accordance with IRC Section R602.12.3, wood structural panel sheathing is required to be attached with 8d nails spaced 4 inches on center at panel ends and edges, and hold downs are required at ends of each continuous length of wood structural panel bracing in accordance with IRC Table R602.12(2).

As illustrated in IRC Figure R602.12 (**FIGURE 10.1**), the capacity of the hold down is determined by the relationship of the ends of braced wall panels between stories. When the ends of braced wall lines are vertically aligned between stories, the cumulative hold down capacity given in IRC Table R602.12(2) is required. The required capacity of these aligned hold downs increases as directed in IRC Table R602.12(2), reproduced in **TABLE 10.2** and illustrated in IRC Figure R602.10 (**FIGURE 10.1**).

When the ends of braced wall lines are not vertically aligned, the single-story hold down capacity given in IRC Table R602.12(2) is used. When the braced wall panel occurs above the first story, the hold-down anchorage must be continued down to the foundation. For example, if a hold down is installed on the second or third story, each story below must have a hold down of the same capacity that is vertically aligned with the hold down on the story above.

> **R602.12.1.4 Minimum length of braced panel.** *Each braced wall panel shall be at least 48 inches in length, covering a minimum of 3 stud spaces where studs are spaced 16 inches on center and covering a minimum of 2 studs where studs are spaced 24 inches on center.*

This code section disallows the use of the effective length provision in IRC Table R602.10.3. All braced wall panels must have a minimum length of 48 inches. This minimum length requirement of 48 inches effectively prohibits the use of continuous sheathing in a manner intended to decrease the panel length to less than 48 inches without the use of hold downs, as expressly stated in IRC Section R602.12.1.6. Note that while wood structural panel sheathing may be used over the whole wall, above and below openings, hold downs only need to be applied to the ends of those elements that are required for intermittent bracing. This is an important consideration when bracing lengths greater than 48 inches are used. For example, if a wall line has a 6-foot length of wood structural panel wall bracing, consisting of one 4-foot and one 2-foot piece attached to a common stud, the hold downs are only required at the start and end of the 6-foot length.

The hold downs are typically required to be nailed into two full-height studs at each end of the braced wall panel element. Designers used to working with the International Building Code (IBC) will recognize that these bracing elements are similar to shear walls.

> **R602.12.1.5 Alternate braced wall panel.** *Alternate braced wall panels described in Section R602.10.3.2 shall not replace the braced wall panel specification of this section.*

The intent of this section is to prevent alternate braced wall panels from being used for the purpose of reducing the required panel length or hold down capacity. The required minimum length of a braced wall panel used in D_0, D_1 and D_2 is 48 inches with the hold down capacity given in IRC Table R602.12(2).

> **R602.12.1.6 Continuously sheathed wall bracing.** *Continuously sheathed provisions of Section R602.10.4 shall not be used in conjunction with the wall bracing provisions of this section.*

Continuous sheathing is prohibited because the capacities of Methods CS-WSP (continuous wood structural panel sheathing), CS-G (wood structural panel adjacent to garage door openings and supporting roof loads only) and CS-PF (continuous portal frame) are considerably lower than fully restrained, 48-inch length panels constructed with 7/16-inch-thick wood structural panel sheathing and attached with 8d common nails at 4 inches on center. While not specifically stated, it can be interpreted that Method CS-SFB (continuous structural fiberboard sheathing) may not be used either.

Summary of stone and masonry veneer provisions

A summary of the code provisions for structures with stone and masonry veneer in SDC A, B and C is provided in ***TABLE 10.3***. A summary of the provisions applicable to structures in SDC D_0, D_1 and D_2 is provided in ***TABLE 10.4***. For convenience, these two tables also include wall bracing adjustment factors for other design considerations, such as story height and braced wall line spacing. Note that the application of stone and masonry veneer is not permitted for townhouses in SDC D_0, D_1 and D_2. The implication is that an engineered design using the IBC is required for townhouses having stone or masonry veneer in these Seismic Design Categories.

The interpretations given in ***TABLE 10.3*** and ***10.4*** assume IRC Section R301.2.2, Exception takes precedence over all other code sections relating to stone and masonry veneer. This exception states that:

> *Detached one- and two-family dwellings located in Seismic Design Category C are exempt from the seismic requirements of the code.*

TABLE 10.3

Stone and masonry veneer requirements for SDC A, B and C

Seismic Design Category	One- and Two-Family Dwellings	Townhouses
A and B	Exempt from all provisions relating to additional bracing requirements for stone and masonry veneer per IRC Section R301.2.2, Exception. Only bracing for wind speed is required.	Exempt from all provisions relating to additional bracing requirements for stone and masonry veneer per IRC Section R301.2.2, Exception. Only bracing for wind speed is required.
C	Exempt from all provisions relating to additional bracing requirements for stone and masonry veneer per IRC Section R301.2.2, Exception. Only bracing for wind speed is required.	Stories in townhouses that support a story above are subject to a bracing length increase per IRC Table R602.12(1)[a]. **Summary of Adjustment Factors:** • Story height per IRC Table R602.10.1.2(3) • Braced wall line spacing factors (of 1.0 to 1.43) per IRC Table R602.10.1.2(3) • Roof/ceiling dead load per IRC Table R602.10.1.2(3) • Cripple wall bracing length and panel spacing per IRC Section R602.10.9 • Bracing length per IRC Table R602.12(1)

a. Wall dead load adjustment factors per IRC Table R602.10.1.2(3) do not apply because this factor would be redundant with the factor in IRC Table R602.12(1) that requires an increased bracing length for the stone and masonry veneer itself.

TABLE 10.4

Stone and masonry veneer requirements for SDC D$_0$, D$_1$ and D$_2$

Story Condition	One- and Two-Family Dwellings	Townhouses
One-story or multi-story with only the bottom story having stone or masonry veneer that does not exceed the first story height. 	Wind and seismic bracing provisions of IRC Sections R602.10 and R602.11 apply. Provisions of R602.12.1 do not apply when stone or masonry veneer exceed the full-height of first story.	While it can be interpreted from IRC Section R602.12.1 that townhouses could be included, the text of the subsequent sections reference a table that is exclusive to one-and two-family dwellings. From an engineering perspective, there is little difference between the impact of a one-story or less stone or masonry wall regardless of the structure type. It is reasonable to assume that as long as the stone or brick veneer wall is the full height or less of the first story, then, like one- and two-family dwellings, no additional bracing is required for townhouses. No guidance is given for townhouses with greater heights of stone or brick veneer and therefore engineering is required.
One, two or three-story with stone or masonry veneer exceeding the first story height[b]	**Summary of adjustment factors[a]:** <ul><li>Story height per IRC Table R602.10.1.2(3)</li><li>Braced wall line spacing per IRC Table R602.10.1.5</li><li>Roof/ceiling dead load per IRC Table R602.10.1.2(3)</li><li>Bracing length per IRC Table R602.12(2)</li></ul> **Additional bracing requirements:** <ul><li>Wall anchorage per IRC Section R602.11.1</li><li>Hold downs per IRC Table R602.12(2)</li><li>Cripple walls not permitted by IRC Section R602.12</li><li>Interior braced wall lines must be supported on continuous foundations</li><li>Braced panel material per IRC Table R602.12(2)</li><li>8' braced panel end distance</li><li>Minimum panel width of 48" per IRC Section R602.12.1.4</li></ul>	Not permitted. The structure is required to be designed in accordance with accepted engineering practice because the provisions of IRC Section R602.12 only permit the application of stone and masonry veneer for one- and two- family dwellings in SDC D$_0$, D$_1$ and D$_2$. As such, the exception to IRC Section R301.2.2.2.5, Item 7 does not apply to townhouses and they must be considered irregular.

a. Wall dead loads per IRC Table R602.10.1.2(3) do not apply because this factor would be redundant with the factor in IRC Table R602.12(1) that requires an increased bracing length for the stone and masonry veneer itself.

b. Three-story structures are not permitted in SDC D$_2$ per IRC Table R602.12(2).

Note that the stone and masonry provisions discussed above only relate to the amount of bracing required for the seismic conditions. Wind bracing in accordance with IRC Table R602.10.1.2(1) must also be considered.

CHAPTER 11

Whole House Considerations

Putting it all together

Previous chapters in this book have explained the bracing code provisions and provided bracing method examples limited to single braced wall lines. This chapter provides application examples of these methods used together on modern house plans. A summary list of the bracing methods is provided in *TABLE 11.1*, while detailed requirements for each method are provided in *CHAPTER 5*.

What is the intent?

When the plan layout of a home design is not specifically addressed by the provisions in the IRC, two questions to ask are:

1. What is the intent of the bracing provisions?

2. Is the intent being met by the bracing provided?

In these situations, it is necessary to communicate with the local building official to determine a proper interpretation. In many cases, the building official may have already encountered the same issue and have a solution.

Note that any concern regarding interpretation of the bracing provisions should be discussed with the building official before the plans are finalized in order to avoid an extensive redesign.

Whole house examples

TABLE 11.1 provides a list of the bracing methods used in this chapter's examples. A more comprehensive table, including a summary of placement and application requirements for each method, is provided at the end of this publication (page 254) for quick reference.

TABLE 11.1

Bracing schedule

Description of bracing methods used in examples. See page 254 for a more comprehensive overview of the bracing methods.

Bracing Method	Method Description
LIB	Let-in-bracing
DWB	Diagonal wood boards
WSP	Wood structural panel
SFB	Structural fiberboard sheathing
GB	Gypsum board
PBS	Particleboard sheathing
PCP	Portland cement plaster
HPS	Hardboard panel siding
ABW	Alternate braced wall
PFH	Intermittent portal frame
PFG	Intermittent portal frame at garage door openings
CS-WSP	Continuous wood structural panel sheathing
CS-G	Wood structural panel adjacent to garage door openings and supporting roof load only
CS-PF	Continuous portal frame
CS-SFB	Continuous structural fiberboard sheathing

Example 11.1 demonstrates how a simple house can comply with the IRC wall bracing provisions. Later examples will demonstrate more complex designs.

Example 11.1: Two-story house in SDC C

The basic wind speed is 90 mph with Exposure Category B. The roof has an eave-to-ridge height of 12 feet. Intermittent Method SFB (structural fiberboard sheathing) will be used as the bracing material on the exterior braced wall lines. Method GB (gypsum board) will be used on the interior braced wall lines. All braced walls have a height of 9 feet.

Example 11.1 highlights:

- The amount of bracing required is based on the ≤ 90 mph wind speed category (IRC Table R602.10.1.2(1)).

- Braced wall panels are permitted to be located a combined distance of 12.5 feet from the ends of braced wall lines (IRC Section R602.10.1.4).

- For a single-family dwelling located in SDC C, the seismic provisions of the IRC do not apply. Therefore, the provisions defining irregular buildings (IRC Section R301.2.2.2.5) and bracing lengths based on SDC (IRC Table R602.10.1.2(2)) do not apply.

- Describes how to compare the amount of bracing required to the amount of bracing provided.

FIGURES 11.1a and *11.1b* show the house and braced wall segments for each story. *TABLES 11.1a* and *11.1b* summarize the amount of bracing required and the amount of bracing provided for each braced wall line on each story.

FIGURE 11.1a

First-story plan with intermittent Method SFB (structural fiberboard sheathing) braced wall panels

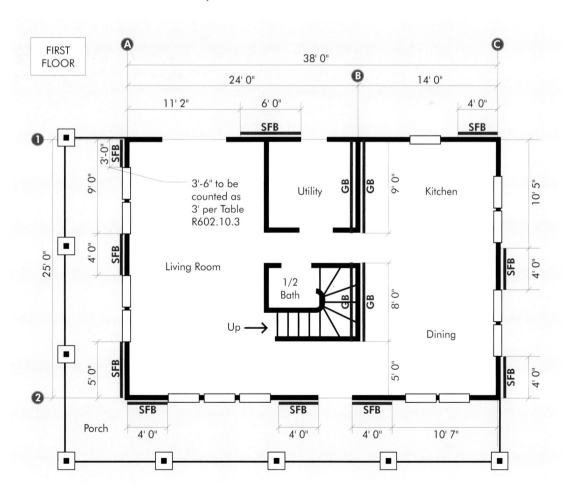

TABLE 11.1a

Calculations for the first of two stories to determine the required bracing length based on wind speed

IRC Table R602.10.1.2(1) Footnotes	Numbered Wall Lines	Lettered Wall Lines
(b) Exposure Category	1.00	1.00
(c) Roof Eave-to-Ridge Height	1.06	1.06
(d) Wall Height	0.95	0.95
(e) Number of Braced Wall Lines	1.00	1.30
Wind Factor Total	**1.01**	**1.31**

Braced Wall Line	Bracing Method	Braced Wall Line Spacing (ft)	Required Bracing (ft)	Wind Factor Total	Total Gypsum Factor	Panels with Hold Downs	Total Required Bracing Length (ft)	Bracing Length Provided (ft)	Status
1	SFB	25.00	9.00	1.01	n/a	n/a	9.09	10.00	Okay
2	SFB	25.00	9.00	1.01	n/a	n/a	9.09	12.00	Okay
A	SFB	24.00	8.70	1.31	n/a	n/a	11.40	12.00	Okay
B	GB w/ 4" nailing	24.00	15.20	1.31	0.70[g]	n/a	13.93	17.00	Okay
C	SFB	14.00	5.40	1.31	n/a	n/a	7.07	8.00	Okay

Footnote g: *…When Method GB braced wall panels installed in accordance with Section R602.10.2 are fastened at 4 inches on center at panel edges, including top and bottom plates, and are blocked at all horizontal joints, multiplying the required bracing percentage for wind loading by 0.7 shall be permitted.* (IRC Table R602.10.1.2(1))

Note that for braced wall line B in Example 11.1, the reduction in required bracing provided for in Footnote g of IRC Table R602.10.1.2(1) was used as an alternative to increasing the length of bracing required along braced wall line B.

FIGURE 11.1b

Second-story plan with intermittent Method SFB (structural fiberboard sheathing) braced wall panels

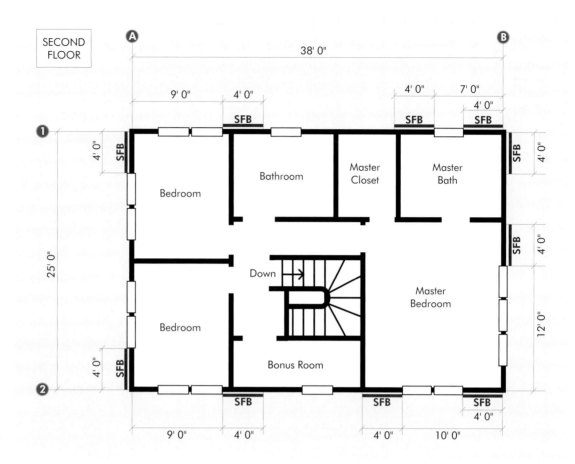

TABLE 11.1b

Calculations for the second of two stories to determine the required bracing length based on wind speed

IRC Table R602.10.1.2(1) Footnotes	Numbered Wall Lines	Lettered Wall Lines
(b) Exposure Category	1.00	1.00
(c) Roof Eave-to-Ridge Height	1.12	1.12
(d) Wall Height	0.95	0.95
(e) Number of Braced Wall Lines	1.00	1.00
Wind Factor Total	**1.06**	**1.06**

Braced Wall Line	Bracing Method	Braced Wall Line Spacing (ft)	Required Bracing (ft)	Wind Factor Total	Total Gypsum Factor	Panels with Hold Downs	Total Required Bracing Length (ft)	Bracing Length Provided (ft)	Status
1	SFB	25.00	4.75	1.06	n/a	n/a	5.04	12.00	Okay
2	SFB	25.00	4.75	1.06	n/a	n/a	5.04	12.00	Okay
A	SFB	38.00	7.10	1.06	n/a	n/a	7.53	8.00	Okay
B	SFB	38.00	7.10	1.06	n/a	n/a	7.53	8.00	Okay

Although Method SFB bracing was used, the solution would be identical if using Method DWB, WSP, PBS, PCP or HPS. Note that if one of the continuous sheathing bracing methods had been used, the required length of bracing would have been less than the required length for the other methods, per IRC Table R602.10.1.2(1).

Example 11.2: Two-story house in SDC D$_2$

The basic wind speed is 85 mph with Exposure Category B. The roof has an eave-to-ridge height of 8 feet. Intermittent Method WSP (wood structural panel) will be used as the bracing material on the exterior braced wall lines. Intermittent Method GB (gypsum board) will be used on the interior braced wall lines. All braced walls have a height of 10 feet.

Example 11.2 highlights:

- The amount of bracing required is based on the greater amount required by the ≤85 mph wind speed category and SDC D$_2$.

- For a single-family dwelling located in SDC D$_0$, D$_1$ and D$_2$, braced wall panels on the second floor cannot be placed over an opening on the first floor except in certain cases (IRC Section R301.2.2.2.5, Item 3, Exceptions).

- In SDC D$_0$, D$_1$ and D$_2$, braced wall panels are not permitted to be located away from the ends of braced wall lines except in certain cases and only with Method WSP (IRC Section R602.10.1.4.1, Exception).

- For structures in SDC D$_0$, D$_1$ and D$_2$, braced wall line spacing is limited to 25 feet (IRC Section R602.10.1.5).

- Describes how to compare the amount of bracing required to the amount of bracing provided.

FIGURES 11.2a and **11.2b** show the house and braced wall segments for each story. **TABLES 11.2a** and **11.2b** summarize the amount of bracing required and the amount of bracing provided for each braced wall line on each story, based on wind and seismic requirements respectively.

FIGURE 11.2a

First-story plan with intermittent Methods WSP (wood structural panel) and GB (gypsum board) braced wall panels

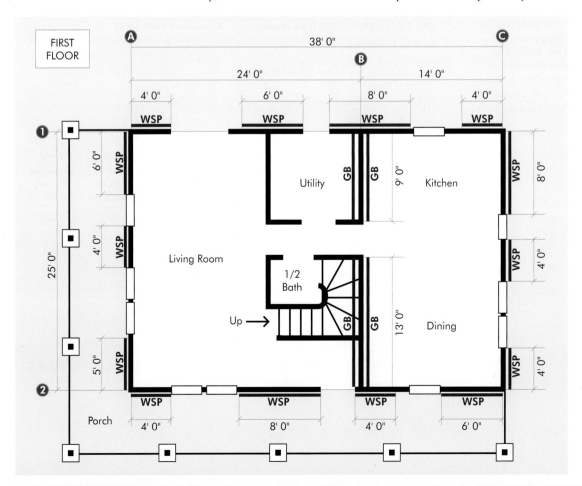

TABLE 11.2a

Calculations for the first of two stories to determine the required bracing length based on wind speed and SDC

Wind Calculations

IRC Table R602.10.1.2(1) Footnotes	Numbered Wall Lines	Lettered Wall Lines
(b) Exposure Category	1.00	1.00
(c) Roof Eave-to-Ridge Height	0.94	0.94
(d) Wall Height	1.00	1.00
(e) Number of Braced Wall Lines	1.00	1.30
Wind Factor Total	**0.94**	**1.22**

Braced Wall Line	Bracing Method	Braced Wall Line Spacing (ft)	Required Bracing (ft)	Wind Factor Total	Total Gypsum Factor	Panels with Hold Downs	Total Required Bracing Length (ft)	Bracing Length Provided (ft)	Status
1	WSP	25.00	8.00	0.94	n/a	n/a	7.52	22.00	Okay
2	WSP	25.00	8.00	0.94	n/a	n/a	7.52	22.00	Okay
A	WSP	24.00	7.70	1.22	n/a	n/a	9.39	15.00	Okay
B	GB	24.00	13.50	1.22	n/a	n/a	16.47	22.00	Okay
C	WSP	14.00	4.70	1.22	n/a	n/a	5.73	16.00	Okay

Seismic Calculations

Seismic Factors (IRC Table R602.10.1.2(3))	Adjustments
Story Height	1.00
Wall Dead Load	1.00
Roof/Ceiling Dead Load	1.00
Stone/Masonry in SDC C-D$_2$	n/a
Cripple Wall	n/a
Seismic Factor Total	**1.00**

Braced Wall Line	Bracing Method	Braced Wall Line Length (ft)	Required Bracing (ft)	Braced Wall Line Spacing Factor	Seismic Factor Total	Total Required Bracing Length (ft)	Bracing Length Provided (ft)	Status
1	WSP	38.00	20.90	1.00	1.00	20.90	22.00	Okay
2	WSP	38.00	20.90	1.00	1.00	20.90	22.00	Okay
A	WSP	25.00	13.75	1.00	1.00	13.75	15.00	Okay
B	GB	25.00	18.75	1.00	1.00	18.75	22.00	Okay
C	WSP	25.00	13.75	1.00	1.00	13.75	16.00	Okay

FIGURE 11.2b

Second-story plan with intermittent Methods WSP (wood structural panel) and GB (gypsum board) braced wall panels

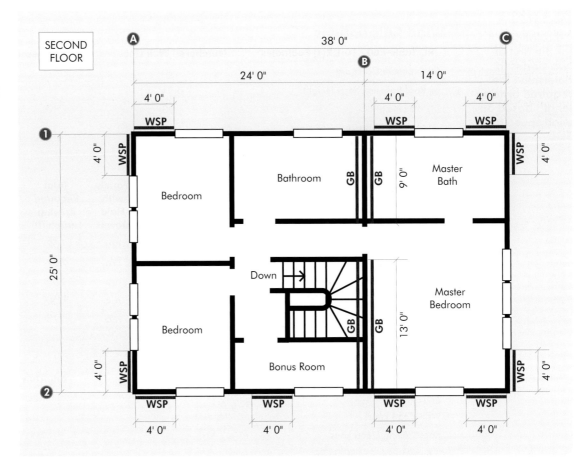

Wind Calculations

IRC Table R602.10.1.2(1) Footnotes	Numbered Wall Lines	Lettered Wall Lines
(b) Exposure Category	1.00	1.00
(c) Roof Eave-to-Ridge Height	0.88	0.88
(d) Wall Height	1.00	1.00
(e) Number of Braced Wall Lines	1.00	1.30
Wind Factor Total	**0.88**	**1.14**

Braced Wall Line	Bracing Method	Braced Wall Line Spacing (ft)	Required Bracing (ft)	Wind Factor Total	Total Gypsum Factor	Panels with Hold Downs	Total Required Bracing Length (ft)	Bracing Length Provided (ft)	Status
1	WSP	25.00	4.25	0.88	n/a	n/a	3.74	12.00	Okay
2	WSP	25.00	4.25	0.88	n/a	n/a	3.74	16.00	Okay
A	WSP	24.00	4.10	1.14	n/a	n/a	4.67	8.00	Okay
B	GB	24.00	7.00	1.14	n/a	n/a	7.98	22.00	Okay
C	WSP	14.00	2.60	1.14	n/a	n/a	2.96	8.00	Okay

Seismic Calculations

Seismic Factors (IRC Table R602.10.1.2(3))	Adjustments
Story Height	1.00
Wall Dead Load	1.00
Roof/Ceiling Dead Load	1.00
Stone/Masonry in SDC C-D$_2$	n/a
Cripple Wall	n/a
Seismic Factor Total	**1.00**

Braced Wall Line	Bracing Method	Braced Wall Line Length (ft)	Required Bracing (ft)	Braced Wall Line Spacing Factor	Seismic Factor Total	Total Required Bracing Length (ft)	Bracing Length Provided (ft)	Status
1	WSP	38.00	9.50	1.00	1.00	9.50	12.00	Okay
2	WSP	38.00	9.50	1.00	1.00	9.50	16.00	Okay
A	WSP	25.00	6.25	1.00	1.00	6.25	8.00	Okay
B	GB	25.00	10.00	1.00	1.00	10.00	22.00	Okay
C	WSP	25.00	6.25	1.00	1.00	6.25	8.00	Okay

Example 11.3: One-story house in SDC A

The basic wind speed is 95 mph with Exposure Category B. The roof has an eave-to-ridge height of 10 feet. Intermittent Methods WSP (wood structural panel), ABW (alternate braced wall), PFG (intermittent portal frame at garage door openings) and GB (gypsum board) will be used as the bracing material on the braced wall lines. All braced walls have a height of 10 feet.

Example 11.3 highlights:

- The amount of bracing required is based on 95 mph wind speed, which is within the ≤100 mph wind speed category.

- Use of the 24-inch-wide minimum Method PFG (intermittent portal frame at garage door openings) panel (IRC Section R602.10.3.4). The actual bracing length of Method PFG is equivalent to its measured length times a factor of 1.5.

- Use of the 32-inch-wide Method ABW (alternate braced wall) panel (IRC Section R602.10.3.2).

- Use of different bracing methods in one wall line.

- Defines the length of braced wall lines using a "box" concept.

- Describes how to compare the amount of bracing required to the amount of bracing provided.

FIGURE 11.3 shows the house and braced wall segments. **TABLE 11.3** summarizes the amount of bracing required and the amount of bracing provided for each braced wall line.

FIGURE 11.3

One-story plan with intermittent Methods WSP (wood structural panel), GB (gypsum board), ABW (alternate braced wall) and PFG (intermittent portal frame at garage door openings) braced wall panels

TABLE 11.3

Calculations to determine the required bracing length based on wind speed

IRC Table R602.10.1.2(1) Footnotes	Numbered Wall Lines	Lettered Wall Lines
(b) Exposure Category	1.00	1.00
(c) Roof Eave-to-Ridge Height	1.00	1.00
(d) Wall Height	1.00	1.00
(e) Number of Braced Wall Lines	1.45	1.60
Wind Factor Total	**1.45**	**1.60**

Braced Wall Line	Bracing Method	Braced Wall Line Spacing (ft)	Required Bracing (ft)	Wind Factor Total	Total Gypsum Factor	Panels with Hold Downs	Total Required Bracing Length (ft)	Bracing Length Provided (ft)	Status
1	WSP, ABW	32.5	7.50	1.45	n/a	n/a	10.88	16.00	Okay
2	WSP, GB	32.5	12.88	1.45	n/a	n/a	18.68	20.00	Okay
3	WSP	20.00	5.00	1.45	n/a	n/a	7.25	8.00	Okay
4	PFG	24.00	5.80	1.45	n/a	n/a	8.41	9.00	Okay
A	GB	16.17	6.97	1.60	2.00[g]	n/a	22.30	25.67	Okay
B	GB	16.17	6.97	1.60	2.00[g]	n/a	22.30	24.00	Okay
C	GB	26.92	10.92	1.60	2.00[g]	n/a	34.94	45.67	Okay
D	GB	26.92	10.92	1.60	2.00[g]	n/a	34.94	36.00	Okay
E	GB	14.58	6.33	1.60	2.00[g]	n/a	20.26	31.33	Okay

Footnote g: *Bracing lengths for Method GB (gypsum board) are based on the application of gypsum board on both faces of a braced wall panel. When Method GB bracing is provided on only one side of the wall, the required bracing amounts shall be doubled...* (IRC Table R602.10.1.2(1))

Note that in the above example, single-sided Method GB bracing was used to facilitate ease of construction. For exterior walls, the use of one-sided Method GB bracing on the interior of the wall gives the designer greater freedom in selecting exterior finishing materials. On interior braced wall lines with single-sided Method GB wall bracing, only the Method GB side of each wall requires the 7-inch on center fastener spacing as required by the bracing method. Where gypsum board is installed as an interior finish material and not as Method GB bracing, the fastener spacing requirements of IRC Tables R602.3(1) or R702.35 are appropriate. Where Method GB is used on both sides of the wall, 7-inch fastener spacing is required on both sides of the wall.

Example 11.4: One-story house in SDC A

The basic wind speed is 85 mph with Exposure Category B. The roof has an eave-to-ridge height of 14 feet. Intermittent braced wall panels constructed from intermittent Methods LIB (let-in bracing), DWB (diagonal wood boards), GB (gypsum board), PBS (particleboard sheathing) and HPS (hardboard panel siding) will be used. All braced walls have a height of 10 feet.

Example 11.4 highlights:

- The amount of bracing required is based on 85 mph winds.

- Use of different bracing methods in one wall line and in different wall lines.

- Defines the length of braced wall lines using a "box" concept.

- Describes how to compare the amount of bracing required to the amount of bracing provided.

FIGURE 11.4 shows the house and braced wall segments. **TABLE 11.4** summarizes the amount of bracing required and the amount of bracing provided for each braced wall line.

FIGURE 11.4

One-story plan with intermittent Methods HPS (hardboard panel siding), GB (gypsum board), DWB (diagonal wood boards), PBS (particleboard sheathing) and LIB (let-in bracing)

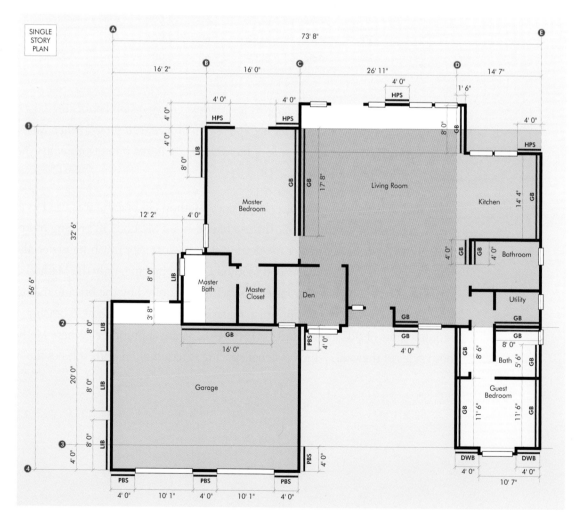

TABLE 11.4

Calculations for the first story to determine the required bracing length based on wind speed

IRC Table R602.10.1.2(1) Footnotes	Numbered Wall Lines	Lettered Wall Lines
(b) Exposure Category	1.00	1.00
(c) Roof Eave-to-Ridge Height	1.24	1.24
(d) Wall Height	1.00	1.00
(e) Number of Braced Wall Lines	1.45	1.60
Wind Factor Total	**1.80**	**1.98**

Braced Wall Line	Bracing Method	Braced Wall Line Spacing (ft)	Required Bracing (ft)	Wind Factor Total	Total Gypsum Factor	Panels with Hold Downs	Total Required Bracing Length (ft)	Bracing Length Provided (ft)	Status
1	HPS	36.33	5.95	1.80	n/a	n/a	10.71	16.00	Okay
2	GB	36.33	10.40	1.80	2.00[g]	n/a	37.44	40.00	Okay
3	DWB	20.00	3.50	1.80	n/a	n/a	6.30	8.00	Okay
4	PBS	24.00	4.10	1.80	n/a	n/a	7.38	12.00	Okay
A	LIB	32.16	9.15	1.98	n/a	n/a	18.12	24.00	Okay
B	LIB	16.16	5.04	1.98	n/a	n/a	9.98	16.00	Okay
C	GB & PBS	26.92	7.73	1.98	n/a	n/a	15.31	25.67	Okay
D	GB	26.92	7.73	1.98	2.00[g]	n/a	30.61	36.00	Okay
E	GB	14.58	4.65	1.98	2.00[g]	n/a	18.41	31.33	Okay

Footnote g: *Bracing lengths for Method GB (gypsum board) are based on the application of gypsum board on both faces of a braced wall panel. When Method GB bracing is provided on only one side of the wall, the required bracing amounts shall be doubled...* (IRC Table R602.10.1.2(1))

Example 11.5: Two-story house in SDC B

The basic wind speed is 105 mph with Exposure Category B. The roof has an eave-to-ridge height of 12 feet. Intermittent Methods WSP (wood structural panel) and ABW (alternate braced wall) will be used as the bracing material on the first story exterior braced wall lines. Method SFB (structural fiberboard sheathing) will be used on the second story exterior braced wall lines. Intermittent Method GB (gypsum board) will be used on the interior braced wall lines for both the first and second stories. All braced walls have a height of 9 feet.

Example 11.5 highlights:

- The amount of bracing required is based on 105 mph winds, which is within the ≤110 mph wind speed category.

- Use of the 32-inch-wide Method ABW (alternate braced wall) panel (IRC Section R602.10.3.2).

- Use of braced wall lines at interior of structure.

- Use of walls greater than 10 feet tall.

- Use of bracing on a second story that does not have bracing on the first story aligned below it.

- Use of different bracing methods in different wall lines and stories.

- Describes how to compare the amount of bracing required to the amount of bracing provided.

The wall at the front entry exceeds the 100 mph wind speed limitations of the allowable stud length IRC Table R602.3.1. IRC Table R602.3(5) may be used to select the proper stud size, provided the limitations of the table are met. In this case, the entrance foyer wall must be a nonbearing wall. Stud heights that fall outside the scope of either of these tables must be sized based on engineering for the gravity and wind or seismic loads imposed.

The maximum allowed height of a braced wall panel is 12 feet. Walls taller than 12 feet are not permitted to contribute to the required bracing. The tall wall in this example may be considered an opening in the braced wall line, because the distance between the inside edges of the adjacent wall bracing panels is less than 21 feet. (If the maximum allowable spacing between the centers of two 4-foot long braced wall panels is 25 feet, than the distance between the inside edges of the braced wall panels is 21 feet. This effectively establishes the maximum distance permitted between intermittent bracing methods.)

FIGURES 11.5a and **11.5b** show the house and braced wall segments. **TABLES 11.5a** and **11.5b** summarize the amount of bracing required and the amount of bracing provided for each braced wall line.

FIGURE 11.5a

First-story plan with intermittent Methods WSP (wood structural panel) and ABW (alternate braced wall) braced wall panels

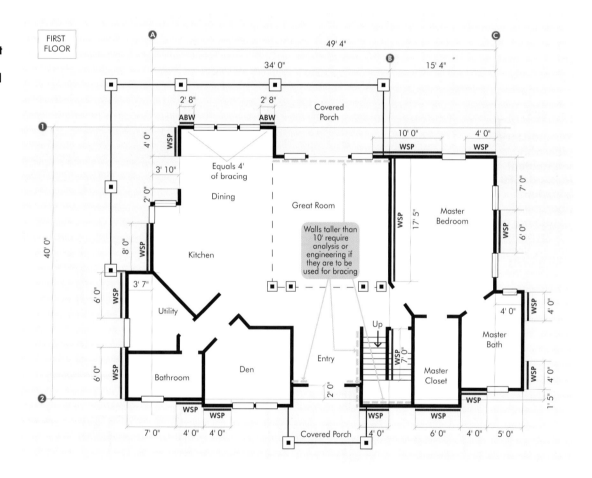

TABLE 11.5a

Calculations for the first of two stories to determine the required bracing length based on wind speed

IRC Table R602.10.1.2(1) Footnotes	Numbered Wall Lines	Lettered Wall Lines
(b) Exposure Category	1.00	1.00
(c) Roof Eave-to-Ridge Height	1.06	1.06
(d) Wall Height	0.95	0.95
(e) Number of Braced Wall Lines	1.00	1.30
Wind Factor Total	**1.01**	**1.31**

Braced Wall Line	Bracing Method	Braced Wall Line Spacing (ft)	Required Bracing (ft)	Wind Factor Total	Total Gypsum Factor	Panels with Hold Downs	Total Required Bracing Length (ft)	Bracing Length Provided (ft)	Status
1	WSP, ABW	40.08	20.54	1.01	n/a	n/a	20.75	22.00	Okay
2	WSP	40.08	20.54	1.01	n/a	n/a	20.75	22.00	Okay
A	WSP	34.00	17.80	1.31	n/a	n/a	23.32	24.00	Okay
B	WSP	34.00	17.80	1.31	n/a	n/a	23.32	24.50	Okay
C	WSP	15.33	8.67	1.31	n/a	n/a	11.36	14.00	Okay

FIGURE 11.5b

Second-story plan with intermittent Methods SFB (structural fiberboard sheathing) and GB (gypsum board) braced wall panels

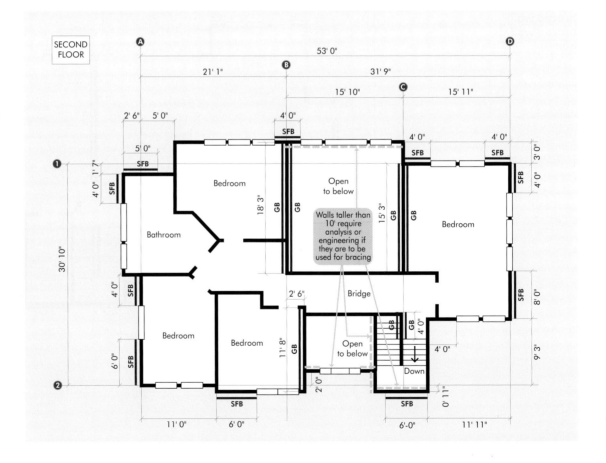

TABLE 11.5b

Calculations for the second of two stories to determine the required bracing length based on wind speed

IRC Table R602.10.1.2(1) Footnotes	Numbered Wall Lines	Lettered Wall Lines
(b) Exposure Category	1.00	1.00
(c) Roof Eave-to-Ridge Height	1.12	1.12
(d) Wall Height	0.95	0.95
(e) Number of Braced Wall Lines	1.00	1.45
Wind Factor Total	**1.06**	**1.54**

Braced Wall Line	Bracing Method	Braced Wall Line Spacing (ft)	Required Bracing (ft)	Wind Factor Total	Total Gypsum Factor	Panels with Hold Downs	Total Required Bracing Length (ft)	Bracing Length Provided (ft)	Status
1	SFB	30.83	8.71	1.06	n/a	n/a	9.23	17	Okay
2	SFB	30.83	8.71	1.06	n/a	n/a	9.23	12	Okay
A	SFB	21.08	6.27	1.54	n/a	n/a	9.66	14	Okay
B	GB	21.08	10.49	1.54	n/a	n/a	16.15	24.08	Okay
C	GB	15.92	8.18	1.54	n/a	n/a	12.57	19	Okay
D	SFB	15.92	4.78	1.54	n/a	n/a	7.36	12	Okay

Example 11.6: One-story house in SDC D$_0$

The basic wind speed is 105 mph with Exposure Category B. The roof has an eave-to-ridge height of 10 feet. Method CS-WSP (continuous wood structural panel sheathing) and CS-PF (continuous portal frame) will be used as the bracing material on the exterior braced wall lines. Intermittent Method GB (gypsum board) will be used on the interior braced wall lines. All braced walls have a height of 8 feet. All windows have clear openings of 64 inches and all doors have clear openings of 80 inches.

Example 11.6 highlights several important issues:

- The amount of bracing required is based on the greater amount required by the ≤110 mph wind speed category and SDC D$_0$.

  ○ Both wind and seismic control the design. For some wall lines, the required bracing length is based on the wind calculation. For other wall lines, the required bracing length is based on the seismic calculation.

- Use of continuous wood structural panel bracing.

- Use of braced wall line spacing factors to accommodate a single room not exceeding 900 square feet per IRC Section R602.10.1.5, Exception.

- Use of different bracing methods in different wall lines and stories.

- Describes how to compare the amount of bracing required to the amount of bracing provided.

FIGURE 11.6 shows the house and braced wall segments. **TABLE 11.6** summarizes the the amount of bracing required and the amount of bracing provided for each braced wall line.

FIGURE 11.6

Single-story plan with Methods CS-WSP (continuous wood structural panel sheathing) and GB (gypsum board) braced wall panels

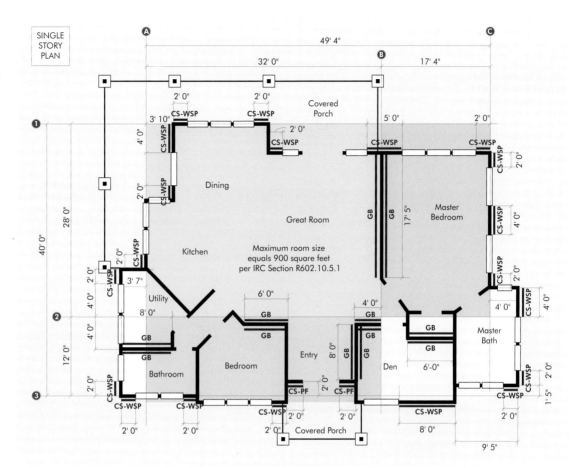

TABLE 11.6

Calculations for the first story to determine the required bracing length based on wind speed and SDC

Wind Calculations

IRC Table R602.10.1.2(1) Footnotes	Numbered Wall Lines	Lettered Wall Lines
(b) Exposure Category	1.00	1.00
(c) Roof Eave-to-Ridge Height	1.00	1.00
(d) Wall Height	0.90	0.90
(e) Number of Braced Wall Lines	1.30	1.30
Wind Factor Total	**1.17**	**1.17**

Braced Wall Line	Bracing Method	Braced Wall Line Spacing (ft)	Required Bracing (ft)	Wind Factor Total	Total Gypsum Factor	Panels with Hold Downs	Total Required Bracing Length (ft)	Bracing Length Provided (ft)	Status
1	CS-WSP	28	6.60	1.17	n/a	n/a	7.72	13	Okay
2	GB	28	13.60	1.17	n/a	n/a	15.91	24	Okay
3	CS-WSP, CS-PF	12	3.40	1.17	n/a	n/a	3.98	20	Okay
A	CS-WSP	32	7.40	1.17	n/a	n/a	8.66	12	Okay
B	GB	32	15.30	1.17	n/a	n/a	17.90	25.5	Okay
C	CS-WSP	17.33	4.47	1.17	n/a	n/a	5.23	14	Okay

Seismic Calculations

Seismic Factors (IRC Table R602.10.1.2(3))	Adjustments
Story Height	1.00
Wall Dead Load	1.00
Roof/Ceiling Dead Load	1.00
Stone/Masonry in SDC C-D$_2$	n/a
Cripple Wall	n/a
Seismic Factor Total	**1.00**

Braced Wall Line	Bracing Method	Braced Wall Line Length (ft)	Required Bracing (ft)	Braced Wall Line Spacing Factor	Seismic Factor Total	Total Required Bracing Length (ft)	Bracing Length Provided (ft)	Status
1	CS-WSP	49.33	8.39	1.12	1.00	9.40	13	Okay
2	GB	49.33	14.80	1.12	1.00	16.58	24	Okay
3	CS-WSP	49.33	8.39	1.00	1.00	8.39	20	Okay
A	CS-WSP	40.0	6.80	1.28	1.00	8.70	12	Okay
B	GB	40.0	12.00	1.28	1.00	15.36	25.5	Okay
C	CS-WSP	40.0	6.80	1.00	1.00	6.80	14	Okay

Notes:

- The distance between wall lines A and B require an adjustment factor.

- The distance between wall lines 1 and 2 require an adjustment factor.

- Wall lines 1, 2, A and B are controlled by the wind calculation. Wall lines 3 and C are controlled by the seismic calculation.

Bracing in high seismic regions

Bracing in high seismic regions requires significantly greater bracing lengths. Additional requirements include:

- Braced wall line spacing must not exceed 25 feet, except that braced wall line spacing can be up to a maximum of 35 feet with the exceptions defined in IRC Section R602.10.1.5.

- Braced wall panels must be located at ends of braced wall lines with an exception for Method WSP (wood structural panel) bracing, as defined in IRC Section R602.10.1.4.1.

- Adhesive attachment of bracing panels is not permitted, as discussed in IRC Section R602.10.2.2.

- Minimum 3-inch x 3-inch x 0.229-inch (9-gage) steel plate washers are required on all anchor bolts for all braced wall line sill plates, as described in IRC Section R602.11.1.

- Additional attachment and foundation requirements apply, as discussed in IRC Section R602.11.

- Houses cannot have irregularities without meeting additional requirements, as discussed in IRC Section R301.2.2.2.5.

These added requirements can restrict the application of prescriptive bracing in high seismic regions. Additionally, the lateral-force-resisting system may require engineering because many modern home designs have architectural features or irregularities which exceed the prescriptive IRC limits.

A good resource for wall bracing applications in high seismic regions is the *Homebuilders Guide to Earthquake Resistant Design and Construction* (FEMA 232).

BRACING T- AND L-SHAPED BUILDINGS

A common problem faced by residential designers is determining the amount and placement of braced wall panels and wall lines in non-rectangular residential structures. Non-rectangular building configurations include T-, L- and U-shaped buildings. Whether the home is fully engineered using the International Building Code (IBC), or designed and constructed by a builder using the prescriptive provisions of the International Residential Code (IRC), it's much easier to apply code provisions to rectangular structures.

So how is the code applied when designing a non-rectangular structure? The answer is the same for both the engineer using the IBC and the designer/builder using the IRC: divide the structure into separate rectangles, determine the shear walls or bracing requirements for each rectangle, and then reconnect the separate rectangles into a unified structure. This method eases the design process while still providing a safe, code-compliant structure. An example is presented below for a L-shaped building. L-shaped buildings are the most common non-rectangular configuration, but the same principles apply to T- and U-shaped buildings, as well as any other shape that can be divided into rectangles. This method can also be used for a large rectangular structure that falls outside of the scope of the IRC, by dividing the structure into two or more elements that do fall within the scope of the code.

Note that in some cases, this method can be conservative. For example, in **FIGURE 11.7**, Box 1 and Box 2 partially shelter each other from the North/South wind, decreasing the overall load on the structure. As a result, bracing along lines 1, 2 and 3 will be somewhat conservative. Exact solutions are often both difficult to arrive at and expensive to build via a fully engineered design, so a little conservatism is economical.

STEP 1: Divide the structure into rectangular elements. There are often multiple ways to do this. Typically, the easiest solution is to divide the building in such a way that the "common side" (or shared side) of the two rectangles contains wall segments which can be used for bracing. See **FIGURE 11.7**:

FIGURE 11.7

Divide structure into rectangular elements

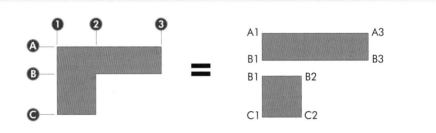

STEP 2: Determine bracing requirements for each individual rectangular element using the IRC bracing provisions. Each individual rectangle is treated and braced as if it were a completely independent, separate structure from the other rectangles. The braced wall line lengths and distance between braced wall lines are measured on each rectangle separately. See **FIGURE 11.8**:

FIGURE 11.8

Determine bracing requirements per the IRC prescriptive provisions for each rectangular element separately

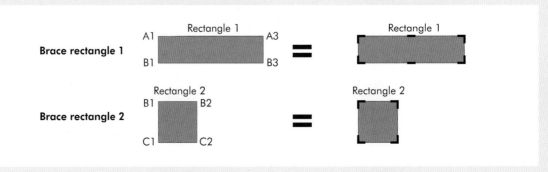

STEP 3: Rejoin the rectangles with bracing provided as shown in *FIGURE 11.9.* The rules that must be applied to the common side when rejoining the rectangles are presented in the list below. Once rejoined, the increased common-side bracing will reflect the appropriate distribution of load. (Note that when this method is used in conjunction with the 2009 IRC Wind Bracing Table (IRC Table R602.10.1.2(1)), Footnote e does not apply.)

FIGURE 11.9

**Rejoin rectangles
with bracing
provided**

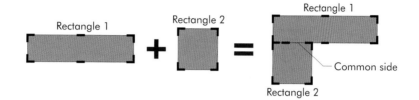

Rules for joining at the common side:

1. The total bracing from both rectangles along the common side must be provided on the common side. In *FIGURE 11.9*, one of the corner bracing elements of the two rectangles was moved to illustrate this point.

2. In the example shown above where the wall line for Rectangle 1 extends past the common side, the entire length of the common braced wall line of Rectangle 1 may be used to position the braced wall panels from both rectangles.

3. The wall bracing location provisions of IRC Section R602.10.1.4 must be met along the common side, as well as along the extended wall line.

4. The 2009 IRC contains provisions permitting mixing of different bracing materials along a single wall line in SDCs A, B and C. This would permit Method GB (gypsum board) bracing along the common side with other panel products used intermittently on the remaining exterior portions of the wall line. These provisions are found in IRC Section R602.10.1.1. Note that when bracing materials are mixed, the following provision applies:

    The length of required bracing for the braced wall line with mixed sheathing types shall have the higher bracing length requirement, in accordance with Tables R602.10.1.2(1) and R602.10.1.2(2), of all types of bracing used.

5. If a physical wall is not available at the common wall location, then all of the bracing for both rectangles must be placed at the exterior extension of the common wall. If the non-existent common wall or an opening in that common wall exceeds 12 feet 6 inches (or 8 feet in SDC D_0, D_1 and D_2) in length, an engineered collector/drag strut (discussed in *CHAPTER 7*) must be used at the common wall location to transfer the bracing from both rectangles into the exterior extension of the common wall. See *FIGURE 11.10*. As an alternative, in Step 1, divide the structure in such a way that there is a physical wall along the common wall. This will provide a location for braced wall panels.

FIGURE 11.10

**No wall at
common wall line**

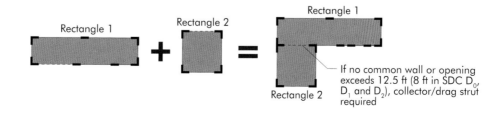

Bibliography

- AAWA. 2003. *Reconnaissance of May 4 and 8, 2003 Kansas-Missouri Tornadoes – A Preliminary Synopsis.* The Wind Engineer – Newsletter of American Association for Wind Engineers, May 2003. American Association for Wind Engineers, Fort Collins, CO. www.aawe.org

- AF&PA. 2001. *Wood Frame Construction Manual.* American Forest & Paper Association, American Wood Council, Washington, DC.

- ANSI/AF&PA. 2008. *Special Design Provisions for Wind and Seismic (SDPWS).* American Forest & Paper Association, American Wood Council, Washington, DC.

- APA – The Engineered Wood Association. 2005. *Missouri Tornadoes – Structural Performance of Wood-Framed Buildings in the Tornadoes of Southwestern Missouri.* Form No. SPE-1118. Tacoma, WA. www.apawood.org

- APA – The Engineered Wood Association. 2005. *Collector Design for Bracing in Conventional Construction.* Form TT-102A. Tacoma, WA. www.apawood.org

- ASCE. 2005. *Minimum Design Loads for Buildings and Other Structures.* ASCE-7-05. American Society of Civil Engineers, Reston, VA.

- BOCA. Through 1999. *National Building Code.* International Code Council, Washington, D.C.

- HUD. 2002. *Framing Connections in Conventional Residential Construction.* U.S. Department of Housing and Urban Development, Washington, DC.

- FEMA. 2006. *Homebuilders Guide to Earthquake Resistant Design and Construction.* FEMA 232. Federal Emergency Management Administration, Jessup, MD.

- IBC. 2006. *International Building Code.* International Code Council, Washington, D.C.

- IBC. 2009. *International Building Code.* International Code Council, Washington, D.C.

- ICBO. Through 1997. *Uniform Building Code.* International Code Council, Washington, D.C.

- ICC/APA – The Engineered Wood Association. 2006. *A Guide to the 2006 IRC Wood Wall Bracing Provisions.* International Code Council, Washington, D.C.

- ICC-400. 2007. *Standard on the Design and Construction of Log Structures.* International Code Council, Washington, D.C.

- ICC-600. 2008. *Standard for Residential Construction in High-Wind Regions.* International Code Council, Washington, D.C.

- IO&TFDC. 1998. *International One- and Two-Family Dwelling Code.* International Code Council, Washington, D.C.

- IRC. 2000. *International Residential Building Code.* International Code Council, Washington, D.C.

- IRC. 2003. *International Residential Building Code.* International Code Council, Washington, D.C.

- IRC. 2006. *International Residential Building Code.* International Code Council, Washington, D.C.

- IRC. 2007. *Supplement to the 2006 International Residential Code.* International Code Council, Washington, D.C.

- IRC. 2009. *International Residential Building Code.* International Code Council, Washington, D.C.

- O&TFDC. 1971 – 1995. *CABO One- and Two-Family Dwelling Code.* Council of American Building Officials, Washington, D.C.

- SBCCI. Through 1999. *Standard Building Code.* International Code Council, Washington, D.C.

- USDA Forest Service. 1999. *Wood Handbook: Wood as an Engineering Material.* USDA Forest Service, Forest Products Laboratory, Madison, WI.

- Zeno Martin, Bryan Readling. 2008. *Midwest Tornadoes – Performance of Wood-Framed Structures.* Wood-Design Focus – A Journal of Contemporary Wood Engineering, Winter 2008. Volume 18, Number 4. Forest Products Society, Madison, WI.

Innovative Building Products

Make sure they are up to code with ICC-ES Evaluation Reports

The ICC-ES Solution

ICC Evaluation Service® (ICC-ES®), a subsidiary of ICC®, was created to assist code officials and industry professionals in verifying that new and innovative building products meet code requirements. This is done through a comprehensive evaluation process that results in the publication of ICC-ES Evaluation Reports for those products that comply with requirements in the code or acceptance critera. Today, more code officials prefer using ICC-ES Evaluation Reports over any other resource to verify products comply with codes.

FREE Access to ICC-ES Evaluation Reports!

ICC-ES Evaluation Report	ESR-4802

Issued March 1, 2008

This report is subject to re-examination in one year.

www.icc-es.org | 1-800-423-6587 | (562) 699-0543   *A Subsidiary of the International Code Council®*

DIVISION: 07—THERMAL AND MOISTURE PROTECTION
Section: 07410—Metal Roof and Wall Panels

REPORT HOLDER:

ACME CUSTOM-BILT PANELS
52380 FLOWER STREET
CHICO, MONTANA 43820
(808) 664-1512
www.custombiltpanels.com

EVALUATION SUBJECT:

CUSTOM-BILT STANDING SEAM METAL ROOF PANELS: CB-150

1.0 EVALUATION SCOPE

Compliance with the following codes:

- 2006 *International Building Code®* (IBC)
- 2006 *International Residential Code®* (IRC)

Properties evaluated:

- Weather resistance
- Fire classification
- Wind uplift resistance

2.0 USES

Custom-Bilt Standing Seam Metal Roof Panels are steel panels complying with IBC Section 1507.4 and IRC Section R905.10. The panels are recognized for use as Class A roof coverings when installed in accordance with this report.

3.0 DESCRIPTION

3.1 Roofing Panels:

Custom-Bilt standing seam roof panels are fabricated in steel and are available in the CB-150 and SL-1750 profiles. The panels are roll-formed at the jobsite to provide the standing seams between panels. See Figures 1 and 3 for panel profiles. The standing seam roof panels are roll-formed from minimum No. 24 gage [0.024 inch thick (0.61 mm)] cold-formed sheet steel. The steel conforms to ASTM A 792, with an aluminum-zinc alloy coating designation of AZ50.

3.2 Decking:

Solid or closely fitted decking must be minimum ¹⁵/₃₂-inch-thick (11.9 mm) wood structural panel or lumber sheathing, complying with IBC Section 2304.7.2 or IRC Section R803, as applicable.

4.0 INSTALLATION

4.1 General:

Installation of the Custom-Bilt Standing Seam Roof Panels must be in accordance with this report, Section 1507.4 of the IBC or Section R905.10 of the IRC, and the manufacturer's published installation instructions. The manufacturer's installation instructions must be available at the jobsite at all times during installation. The roof panels must be installed on solid or closely fitted decking, as specified in Section 3.2. Accessories such as gutters, drip angles, fascias, ridge caps, window or gable trim, valley and hip flashings, etc., are fabricated to suit each job condition. Details must be submitted to the code official for each installation.

4.2 Roof Panel Installation:

4.2.1 CB-150: The CB-150 roof panels are installed on roof shaving a minimum slope of 2:12 (17 percent). The roof panels are installed over the optional underlayment and secured to the sheathing with the panel clip. The clips are located at each panel rib side lap spaced 6 inches (152 mm) from all ends and at a maximum of 4 feet (1.22 m) on center along the length of the rib, and fastened with a minimum of two No. 10 by 1-inch pan head corrosion-resistant screws. The panel ribs are mechanically seamed twice, each pass at 90 degrees, resulting in a double-locking fold.

4.3 Fire Classification:

The steel panels are considered Class A roof coverings in accordance with the exception to IBC Section 1505.2 and IRC Section R902.1.

4.4 Wind Uplift Resistance:

The systems described in Section 3.0 and installed in accordance with Sections 4.1 and 4.2 have an allowable wind uplift resistance of 45 pounds per square foot (2.15 kPa).

5.0 CONDITIONS OF USE

The standing seam metal roof panels described in this report comply with, or are suitable alternatives to what is specified in, those codes listed in Section 1.0 of this report, subject to the following conditions:

5.1 Installation must comply with this report, the applicable code, and the manufacturer's published installation instructions. If there is a conflict between this report and the manufacturer's published installation instructions, this report governs.

5.2 The required design wind loads must be determined for each project. Wind uplift pressure on any roof area must not exceed 45 pounds per square foot (2.15 kPa).

6.0 EVIDENCE SUBMITTED

Data in accordance with the ICC-ES Acceptance Criteria for Metal Roof Coverings (AC166), dated October 2007.

7.0 IDENTIFICTION

Each standing seam metal roof panel is identified with a label bearing the product name, the material type and gage, the Acme Custom-Bilt Panels name and address, and the evaluation report number (ESR-4802).

William Gregory
Building and Plumbing Inspector
Town of Yorktown, New York

"We've been using ICC-ES Evaluation Reports as a basis of product approval since 2002. I would recommend them to any jurisdiction building department, particularly in light of the many new products that regularly move into the market. It's good to have a group like ICC-ES evaluating these products with a consistent and reliable methodology that we can trust."

Becky Baker, CBO
Director/Building Official
Jefferson County, Colorado

"The ICC-ES Evaluation Reports are designed with the end user in mind to help determine if building products comply with code. The reports are easily accessible, and the information is in a format that is useable by plans examiners and inspectors as well as design professionals and contractors."

VIEW ICC-ES EVALUATION REPORTS ONLINE!

www.icc-es.org

Don't Miss Out On Valuable ICC Membership Benefits. Join ICC Today!

Join the largest and most respected building code and safety organization. As an official member of the International Code Council®, these great ICC® benefits are at your fingertips.

EXCLUSIVE MEMBER DISCOUNTS

ICC members enjoy exclusive discounts on codes, technical publications, seminars, plan reviews, educational materials, videos, and other products and services.

TECHNICAL SUPPORT

ICC members get expert code support services, opinions, and technical assistance from experienced engineers and architects, backed by the world's leading repository of code publications.

FREE CODE—LATEST EDITION

Most new individual members receive a free code from the latest edition of the International Codes®. New corporate and governmental members receive one set of major International Codes (Building, Residential, Fire, Fuel Gas, Mechanical, Plumbing, Private Sewage Disposal).

FREE CODE MONOGRAPHS

Code monographs and other materials on proposed International Code revisions are provided free to ICC members upon request.

PROFESSIONAL DEVELOPMENT

Receive Member Discounts for on-site training, institutes, symposiums, audio virtual seminars, and on-line training! ICC delivers educational programs that enable members to transition to the I-Codes®, interpret and enforce codes, perform plan reviews, design and build safe structures, and perform administrative functions more effectively and with greater efficiency. Members also enjoy special educational offerings that provide a forum to learn about and discuss current and emerging issues that affect the building industry.

ENHANCE YOUR CAREER

ICC keeps you current on the latest building codes, methods, and materials. Our conferences, job postings, and educational programs can also help you advance your career.

CODE NEWS

ICC members have the inside track for code news and industry updates via e-mails, newsletters, conferences, chapter meetings, networking, and the ICC website (www.iccsafe.org). Obtain code opinions, reports, adoption updates, and more. Without exception, ICC is your number one source for the very latest code and safety standards information.

MEMBER RECOGNITION

Improve your standing and prestige among your peers. ICC member cards, wall certificates, and logo decals identify your commitment to the community and to the safety of people worldwide.

ICC NETWORKING

Take advantage of exciting new opportunities to network with colleagues, future employers, potential business partners, industry experts, and more than 50,000 ICC members. ICC also has over 300 chapters across North America and around the globe to help you stay informed on local events, to consult with other professionals, and to enhance your reputation in the local community.

JOIN NOW! 1-888-422-7233, x33804 | www.iccsafe.org/membership

ICC INTERNATIONAL CODE COUNCIL®

People Helping People Build a Safer World™

09-01530

APA Delivers Confidence.

When you see fully sheathed wood walls with APA's stamp, you can be confident the home is built with the materials required to meet the most stringent bracing methods. Since the early fifties, APA has collaborated with code bodies to improve wall performance. This collaboration has included a significant evolution in the use of prescriptive wall bracing techniques specifying plywood and OSB. When it comes to experience, our quality, technical and field

▼ Members of APA's Technical Services Division pose with their newest test equipment. This device applies a 32-ton tension load to a full scale, three-dimensional "house". From the test data, APA is able to determine how code compliant wall bracing and perpendicular wall lines behave in high wind or seismic events.

APA Full-Scale House Test Project

services personnel have over 1,100 combined years in the industry. In high wind and seismic events, APA's expertise is on display with products that meet the industry's highest performance standards and wall bracing applications aimed at reducing the risk of catastrophic home failure. For our latest research in building strong, safe and durable structures, visit **www.apawood.org** or **www.wallbracing.org**.

▲ This test simulates seismic lateral loads over dozens of cycles to the same wall line. APA researchers can then determine the suitability of various bracing details for seismic performance.

▲ This panel reaches failure point for flat-wise bending, one of 50,000 samples tested each year by APA's Quality Services Division. Along with regular mill audits, these tests help ensure that APA-stamped products meet rigorous performance criteria.

▲ APA construction details and building recommendations are available online at www.apawood.org. Here you will find over 200 CAD details, along with Evaluation Reports, Technical Topics and Builder Tips. The construction spectrum ranges from residential to commercial and specialty applications, such as weather-resistive and engineered systems.

▲ APA's Field Services Division is your partner on site. In every major jurisdiction, we offer free training and consultation services on the proper specification and application of engineered wood systems.

Need an immediate solution?
Contact APA's Product Support Help Desk at (253) 620-7400.

2009 IRC Code Reference Index

2009 IRC Bracing Methods Overview

Bracing Method	Description[a]	Braced Wall Panel Widths		Max. Height	Max. Distance of First Braced Wall Panel From End of Braced Wall Line	
		Wall Height	Min. Width		SDC A, B & C Combined Total From Both Ends	SDC D_0, D_1 & D_2 Combined Total From Both Ends
LIB	Let-in bracing[b] IRC Table R602.10.2	8' 9' 10' 11' 12'	4' 7" to 8' 5' 2" to 9' 5' 9" to 10' 6' 4" to 11' 6' 11" to 12'	12'	12.5'	Not permitted
DWB	Diagonal wood boards IRC Table R602.10.2					
SFB	Structural fiberboard sheathing IRC Table R602.10.2	8' 9' 10' 11' 12'	36"[c] to 4' 0" 42"[c] to 4' 0" 4' 0" 4' 5" 4 10"	12'	12.5'	0
PBS	Particleboard sheathing IRC Table R602.10.2					
PCP	Portland cement plaster IRC Table R602.10.2					
HPS	Hardboard panel siding IRC Table R602.10.2					
WSP	Wood structural panel IRC Table R602.10.2	8' 9' 10' 11' 12'	36"[c] to 4' 0" 42"[c] to 4' 0" 4' 0" 4' 5' 4 10"	12'	12.5'	8' with 1800 lb. hold down, or a minimum 24" WSP panel each side of corner required
GB	Gypsum board – Single-sided with finish material IRC Table R602.10.2	8' 9' 10' 11' 12'	8' 0" 8' 0" 8' 0" 8' 10" 9' 8"	12'	12.5'	0
	Gypsum board – Double-sided IRC Table R602.10.2	8' 9' 10' 11' 12'	4' 0" 4' 0" 4' 0" 4' 5" 4' 10"			
ABW	Alternate braced wall (SDC A, B & C) IRC Section R602.10.3.2	8' 9' 10' 11' 12'	28" 32" 34" 38" 42"	12'	12.5'	8'
	Alternate braced wall (SDC D_0, D_1 & D_2) IRC Section R602.10.3.2	8' 9' 10' 11' 12'	32" 32" 34" Not permitted Not permitted	10'	12.5'	8'

Bracing Method	Description[a]	Braced Wall Panel Widths		Max. Height	Max. Distance of First Braced Wall Panel From End of Braced Wall Line	
		Wall Height	Min. Width		SDC A, B & C Combined Total From Both Ends	SDC D_0, D_1 & D_2 Combined Total From Both Ends
PFG	Intermittent portal frame at garage door openings in SDC A, B & C IRC Section R602.10.3.4	8' 9' 10' 11' 12'	24" 27" 30" Not permitted Not permitted	10'[d]	0'	Not permitted
PFH	Intermittent portal frame[e] IRC Section R602.10.3.3	8'-10' 11' 12'	16" for one story. 24" for first of two stories. Not permitted Not permitted	10'[d]	12.5'	8'
CS-WSP	Continuous wood structural panel sheathing IRC Table R602.10.4.1	See IRC Table R602.10.4.2[f]. Braced wall panel length varies with wall height and adjacent opening height.		12'	25' with 12.5' max. per wall end 24" min. corner detail required	8', 24" corner detail or hold down required
CS-G	Continuous sheathing - wood structural panel adjacent to garage door openings and supporting roof loads only IRC Table R602.10.4.1	8' 9' 10' 11' 12'	24" 27" 30" 33"[f] 36"[f]	12'[d]	0'	0'
CS-PF	Continuous portal frame IRC Table R602.10.4.1	8' 9' 10' 11' 12'	16" 18" 20" 22"[f] 24"[f]	12'[d]	25' with 12.5' max. per wall end 24" min. corner detail required	8', 24" corner detail or hold down required
CS-SFB	Continuous structural fiberboard sheathing[g] IRC Section R602.10.5	See IRC Table R602.10.5(2)[h]. Braced wall panel length varies with wall height and adjacent opening height.		12'	25' with 12.5' max. per wall end 32" min. corner detail required	Not Permitted

a. Braced wall panels shall be located not more than 25 ft on center.

b. Not permitted on the first of three stories for wind. Only permitted for a one-story structure in SDC C. Not permitted in SDC D_0, D_1 or D_2. See IRC Tables R602.10.1.2(1) and R602.10.1.2(2) (**TABLE 6.1** and **6.2** of this publication).

c. See IRC Section R602.10.3 and Table R602.10.3 (**TABLE 5.3** of this publication) for effective bracing length.

d. Clear opening height = 10 ft minus depth of header.

e. Not permitted on the first of three stories or any story above the first story. Maximum wall height is 10 ft.

f. For expanded table, see **TABLE 5.6** of this publication.

g. When continuous structural fiberboard sheathing (Method CS-SFB) is used where the basic wind speed is in excess of 100 mph or the SDC is D_0, D_1 or D_2, the CS-SFB braced wall line is required to be designed in accordance with accepted engineering practice and the provisions of the International Building Code (IBC). See IRC Section R602.10.5.4.

h. See **TABLE 5.9** of this publication.